Numerical Modeling of COVID-19 Neurological Effects

Numerical Modeling of COVID-19 Neurological Effects
ODE/PDE Analysis in R

William E. Schiesser

CRC Press
Taylor & Francis Group
Boca Raton London New York

CRC Press is an imprint of the
Taylor & Francis Group, an **informa** business

First edition published 2022
by CRC Press
6000 Broken Sound Parkway NW, Suite 300, Boca Raton, FL 33487-2742

and by CRC Press
2 Park Square, Milton Park, Abingdon, Oxon, OX14 4RN

© 2022 William E. Schiesser

CRC Press is an imprint of Taylor & Francis Group, LLC

Reasonable efforts have been made to publish reliable data and information, but the author and publisher cannot assume responsibility for the validity of all materials or the consequences of their use. The authors and publishers have attempted to trace the copyright holders of all material reproduced in this publication and apologize to copyright holders if permission to publish in this form has not been obtained. If any copyright material has not been acknowledged please write and let us know so we may rectify in any future reprint.

Except as permitted under U.S. Copyright Law, no part of this book may be reprinted, reproduced, transmitted, or utilized in any form by any electronic, mechanical, or other means, now known or hereafter invented, including photocopying, microfilming, and recording, or in any information storage or retrieval system, without written permission from the publishers.

For permission to photocopy or use material electronically from this work, access www.copyright.com or contact the Copyright Clearance Center, Inc. (CCC), 222 Rosewood Drive, Danvers, MA 01923, 978-750-8400. For works that are not available on CCC please contact mpkbookspermissions@tandf.co.uk

Trademark notice: Product or corporate names may be trademarks or registered trademarks and are used only for identification and explanation without intent to infringe.

ISBN: 978-1-032-15211-0 (hbk)
ISBN: 978-1-032-15213-4 (pbk)
ISBN: 978-1-003-24305-2 (ebk)

DOI: 10.1201/9781003243052

Publisher's note: This book has been prepared from camera-ready copy provided by the authors.

Access the Support Material: http://www.lehigh.edu/~wes1/cne_download

Typeset in Times New Roman
by codeMantra

Contents

Preface vii

Chapter 1   Source of Neurological Effects One PDE Model 1

        1.1   Introduction 1
               1.1.1   One PDE model formulation 1
               1.1.2   Summary and conclusions 2
        References 3

Chapter 2   Implementation of the One PDE Model 5

        2.1   Introduction 5
               2.1.1   R routines for the one PDE model 5
               2.1.2   Summary and conclusions 15
        References 16

Chapter 3   Two PDE Model 17

        3.1   Introduction 17
               3.1.1   Two PDE model formulation 17
               3.1.2   Summary and conclusions 18
        Reference 19

Chapter 4   Implementation of the Two PDE Model 21

        4.1   Introduction 21
               4.1.1   R routines for the two PDE model 21
               4.1.2   Summary and conclusions 38
        References 38

Chapter 5   Three PDE Model 39

        5.1   Introduction 39
               5.1.1   Three PDE model formulation 39
               5.1.2   Three PDE model implementation 39
               5.1.3   Summary and conclusions 53
        Reference 53

Chapter 6   Case Studies 55

        6.1   Introduction 55
               6.1.1   Time variation of the brain O_2 concentration 55
               6.1.2   LHS PDE time derivatives 68

	6.1.3	Analysis of PDE RHS terms	90
	6.1.4	Summary and conclusions	98
Reference			98

Appendix A: Introduction to PDE Analysis 99

Index 183

Preface

Covid-19 is primarily a respiratory disease which results in impaired oxygenation of blood. The O_2-deficient blood then moves through the body, and for the study in this book, the focus is on the blood flowing to the brain. The dynamics of blood flow along the brain capillaries and tissue is modeled as systems of ordinary and partial differential equations (ODE/PDEs).

As further background (from [3]):

While fewer people who get COVID are dying, not all of them are recovering. We don't know how many people will remain hobbled, long-term. But it is plausible that tens of thousands in the United States may never be the same again.

Additional background is provided in [1, 2, 4, 5].

The ODE/PDE methodology is presented through a series of examples,

1. A basic one PDE model for O_2 concentration in the brain capillary blood.
2. A two PDE model for O_2 concentration in the brain capillary blood and in the brain tissue, with O_2 transport across the blood brain barrier (BBB).
3. The two model extended to three PDEs to include the brain functional neuron cell density.

Cognitive impairment could result from reduced neuron cell density in time and space (in the brain) that follows from lowered O_2 concentration (hypoxia).

The computer-based implementation of the example models is presented through routines coded (programmed) in R, a quality, open-source scientific computing system that is readily available from the Internet. Formal mathematics is minimized, e.g., no theorems and proofs. Rather, the presentation is given through detailed examples that the reader/researcher/analyst can execute on modest computers. The PDE analysis is based on the method of lines (MOL), an established general algorithm for PDEs, implemented with finite differences. Appendix A is a basic introduction to PDE modeling and computer-based analysis the reader can consult when starting on Chapters 1–6.

The routines are available from a download link so that the example models can be executed without having to first study numerical methods and computer coding. The routines can then be applied to variations and extensions of the blood/brain hypoxia models, such as changes in the ODE/PDE parameters (constants) and form of the model equations.

The author would welcome comments/suggestions concerning this approach to the analysis of brain hypoxia resulting from Covid-19.

W. E. Schiesser
Bethlehem, PA, USA

REFERENCES

1. Bridwell, R., B. Long and M. Gottlieb (2020), Neuorologic complications of COVID-19, *American Journal of Emergency Medicine*, 38, pp 1549.e3–1549.e7.
2. Collins, F. (2021), Taking a closer look at COVID-19's effects on the brain, NIH Director's Blog, January 14, 2021.
3. https://www.health.harvard.edu/diseases-and-conditions/what-are-the-long-lasting-effects-of-covid-19.
4. Nuzzo, D., and P. Picone (2020), Potential neurological effects of severe COVID-19 infection, *Neuroscience Research*, 158, pp 1–5.
5. Wood, H. (2020), New insights into the neurological effects of COVID-19, *Nature Reviews Neurology*, June 26, 2020.

1 Source of Neurological Effects One PDE Model

1.1 INTRODUCTION

COVID-19 neurological effects are generally described in the following introductory statement [2]:

> While primarily a respiratory disease, COVID-19 can also lead to neurological problems. The first of these symptoms might be the loss of smell and taste, while some people also may later battle headaches, debilitating fatigue, and trouble thinking clearly, sometimes referred to as "brain fog". All of these symptoms have researchers wondering how exactly the coronavirus that causes COVID-19, SARS-CoV-2, affects the human brain.

This book focuses on a mathematical model describing a reduction in oxygen (O_2) to the brain resulting from impaired respiratory function of the lungs. The reduced brain O_2 (hypoxia) decreases the population density of normal (healthy) neuron cells that then leads to brain cognitive impairment.

Patients who experience diminished O_2 in the capillary blood to the brain resulting from respiratory impairment are sometimes termed "Covid long haulers", long haul Covids", or "long Covids".

Additional background concerning COVID-19 neurological effects is given in [1, 3, 4].

1.1.1 ONE PDE MODEL FORMULATION

A mass balance on the O_2 concentration in the capillary blood flowing in the brain follows (the balance is derived in chapter appendix A1):

$$\frac{\partial u_1}{\partial t} = -v_z \frac{\partial u_1}{\partial z} - (2/r_l)k_{m1}(u_1 - u_{2n}) \qquad (1.1\text{-}1)$$

Equation (1.1-1) states that the time rate of change of the blood O_2 concentration $\left(\dfrac{\partial u_1}{\partial t}\right)$ equals the sum of the rate of convection of O_2 along the capillary $\left(-v_z \dfrac{\partial u_1}{\partial z}\right)$ and the rate of transfer of O_2 across the BBB $(-(2/r_l)k_{m1}(u_1 - u_{2n}))$.

DOI: 10.1201/9781003243052-1

Table 1.1
PDE model variables, parameters

$u_1(z,t)$	oxygen concentration in capillary blood
z	distance along the blood capillary and the brain tissue
r_l	radius of the blood capillary (and inner radius of the blood brain barrier, (BBB))
t	time
v_z	superficial blood flow velocity
k_{m1}	coefficient for mass transfer of O_2 across the BBB
u_{2n}	normalized brain tissue O_2 concentration

Equation (1.1-1) is first order in t so that it requires one initial condition (IC).

$$u_1(z, t=0) = u_{1n} \tag{1.1-2}$$

u_{1n} is the normalized blood O_2 concentration used as an IC.

Equation (1.1-1) is first order in z and requires one boundary condition (BC).

$$u_1(z=z_l, t) = u_{1e}(t) \tag{1.1-3}$$

Equations (1.1) constitute the one PDE model that is implemented in R[1] as explained in Chapter 2.

1.1.2 SUMMARY AND CONCLUSIONS

The one PDE model for O_2 entering the brain through capillary blood flow consists of eqs. (1.1). These equations are a test case with constant O_2 concentration in the brain tissue adjacent to the BBB, u_{2n}. Equations (1.1) are implemented in R in Chapter 2. Variations in the brain tissue O_2 concentration are considered in subsequent chapters through the addition of PDEs.

APPENDIX A1: DERIVATION OF THE BLOOD O_2 BALANCE

A mass balance on the blood with the incremental volume $\pi r_l^2 \Delta z$ (r_l is the radius of the capillary) gives

$$\pi r_l^2 \Delta z \frac{\partial u_1(z,t)}{\partial t} = \pi r_l^2 v_z(t) u_1(z,t)|_z - \pi r_l^2 v_z(t) u_1(z,t)|_{z+\Delta z}$$
$$- 2\pi r_l \Delta z k_{m1}(u_1(z,t) - u_{n2}) \tag{A1-1}$$

The terms in eq. (A1-1) are

---

[1]R is a quality, open source scientific programming system that is available for download through an Internet connection, http://www.r-project.org/, http://cran.fhcrc.org/.

- $\pi r_1^2 \Delta z \dfrac{\partial u_1(z,t)}{\partial t}$: accumulation of O_2 in the incremental volume $\pi r_1^2 \Delta z$. If this term is negative (from the sum of the RHS terms), the O_2 concentration decreases (is depleted) with time.
- $\pi r_1^2 v_z(z,t) u_1(z,t)|_z$: flow (convection) at velocity v_z[2] of the O_2 transported into the incremental volume at z ($v_z > 0$).
- $-\pi r_1^2 v_z(t) u_1(z,t)|_{z+\Delta z}$: flow (convection) at velocity v_z of the O_2 transported out of the incremental volume at $z+\Delta z$ ($v_z > 0$).
- $-2\pi r_1 \Delta z k_{m1}(u_1(z,t) - u_{n2})$: rate of mass transfer of O_2 across the BBB. The transfer area is $2\pi r_1 \Delta z$ and the mass transfer coefficient (BBB permeability) is k_{m1}.

If this term is negative, the transfer is from the blood through the BBB into the brain tissue, which reduces $\dfrac{\partial u_1(z,t)}{\partial t}$

If eq. (A1-1) is divided by the incremental volume $\pi r_1^2 \Delta z$ (the coefficient of $\dfrac{\partial u_1(z,t)}{\partial t}$), after minor rearrangement,

$$\dfrac{\partial u_1(z,t)}{\partial t} = -\dfrac{v_z(t) u_1(z,t)|_{z+\Delta z} - v_z(t) u_1(z,t)|_z}{\Delta z} - \dfrac{2}{r_1} k_{m1}(u_1(z,t) - u_{n2}) \tag{A1-2}$$

For $\Delta z \to 0$, eq. (A1-2) becomes a PDE

$$\dfrac{\partial u_1(z,t)}{\partial t} = -\dfrac{\partial v_z(t) u_1(z,t)}{\partial z} - \dfrac{2}{r_1} k_{m1}(u_1(z,t) - u_{n2}) \tag{A1-3}$$

Equation (A1-3) is eq. (1.1-1) with $v_z(t) = v_z$ (a constant).

REFERENCES

1. Bridwell, R., B. Long and M. Gottlieb (2020), Neuorologic complications of COVID-19, *American Journal of Emergency Medicine*, **38**, pp. 1549.e3–1549.e7.
2. Collins, F. (2021), Taking a closer look at COVID-19's effects on the brain, *NIH Director's Blog*, January 14, 2021.
3. Nuzzo, D. and P. Picone (2020), Potential neurological effects of severe COVID-19 infection, *Neuroscience Research*, **158**, pp. 1–5.
4. Wood, H. (2020), New insights into the neurological effects of COVID-19, *Nature Reviews Neurology*, June 26, 2020.

---

[2] v_z is assumed to be uniform (constant) across the capillary. This is an idealization since the velocity will have a radial profile with zero velocity at the capillary wall (the interface between the blood and the BBB inner surface) and a maximum velocity at the capillary centerline. v_z is termed a *superficial velocity* with the property that multiplication by the capillary cross sectional area πr_1^2 gives the blood volumetric flow rate in the capillary.

2 Implementation of the One PDE Model

2.1 INTRODUCTION

The one PDE model of Chapter 1 for the O_2 in the capillary blood (eqs. (1.1)) is implemented in R routines discussed in this chapter.

2.1.1 R ROUTINES FOR THE ONE PDE MODEL

Equations (1.1) that constitute the PDE model for the O_2 concentration in the capillary blood are implemented with the following R routines, starting with a main program.

Main program for O_2 concentration

The main program for eqs. (1.1) follows.

```
#
#  One PDE model
#
# Delete previous workspaces
  rm(list=ls(all=TRUE))
#
# Access ODE integrator
  library("deSolve");
#
# Access functions for numerical solution
  setwd("f:/brain hypoxia/chap2");
  source("pde1a.R");
#
# Parameters
  nz=21;
  rl=1;
  km1=0;
  km1=1;
  vz=1;
  u1e=0.75;
  u1n=1;
  u2n=0;
#
# Spatial grid in z
```

```
  zl=0;zu=1;dz=(zu-zl)/(nz-1);
  z=seq(from=zl,to=zu,by=dz);
#
# Independent variable for ODE integration
  t0=0;tf=1;nout=21;
  tout=seq(from=t0,to=tf,by=(tf-t0)/(nout-1));
#
# Initial condition (t=0)
  u0=rep(u1n,nz);
  ncall=0;
#
# ODE integration
  out=lsodes(y=u0,times=tout,func=pde1a,
      sparsetype="sparseint",rtol=1e-6,
      atol=1e-6,maxord=5);
  nrow(out)
  ncol(out)
#
# Arrays for plotting numerical solution
  u1=matrix(0,nrow=nz,ncol=nout);
  for(it in 1:nout){
    for(iz in 1:nz){
      u1[iz,it]=out[it,iz+1];
    }
    u1[1,it]=u1e;
  }
#
# Display numerical solution
  iv=seq(from=1,to=nout,by=4);
  for(it in iv){
    cat(sprintf("\n    t       z     u1(z,t)\n"));
    iv=seq(from=1,to=nz,by=2);
    for(iz in iv){
      cat(sprintf("%6.2f%6.2f%12.3e\n",
          tout[it],z[iz],u1[iz,it]));
    }
  }
#
# Calls to ODE routine
  cat(sprintf("\n\n ncall = %5d\n\n",ncall));
#
# Plot PDE solution
#
# u1
```

Implementation of the One PDE Model

```
  par(mfrow=c(1,1));
  matplot(x=z[2:nz],y=u1[2:nz,],type="l",xlab="z",ylab="u1(z,t)",
          xlim=c(zl,zu),lty=1,main="",lwd=2,col="black");
  persp(z,tout,u1,theta=60,phi=45,
        xlim=c(zl,zu),ylim=c(t0,tf),zlim=c(0,1.1),
        xlab="z",ylab="t",zlab="u1(z,t)");
```

Listing 2.1: Main program for eqs. (1.1)

We can note the following details about Listing 2.1.

- Previous workspaces are deleted.

    ```
    #
    #   One PDE model
    #
    # Delete previous workspaces
      rm(list=ls(all=TRUE))
    ```

- The R ODE integrator library deSolve is accessed [2].

    ```
    #
    # Access ODE integrator
      library("deSolve");
    #
    # Access functions for numerical solution
      setwd("f:/brain hypoxia/chap2");
      source("pde1a.R");
    ```

    Then the directory with the files for the solution of eqs. (1.1) is designated. Note that setwd (set working directory) uses / rather than the usual \.
- The model parameters are specified numerically.

    ```
    #
    # Parameters
      nz=21;
      rl=1;
      km1=0;
      km1=1;
      vz=1;
      u1e=0.75;
      u1n=1;
      u2n=0;
    ```

where

- nz: number of spatial grid points for eq. (1.1-1).
- rl: radius of the blood capillary (and inner radius of the blood brain barrier (BBB))
- km1: mass transfer (permeability) coefficient for O_2 across the BBB in eq. (1.1-1). $k_m = 0$ was used initially during the development of the coding (programming). The mass transfer coefficient was then changed to $k_m = 1$.
- vz: capillary blood flow superficial velocity in eq. (1.1-1).
- u1e: entering blood O_2 concentration in BC (1.1-3).
- u1n: normalized blood O_2 concentration in IC (1.1-2).
- u2n: normalized brain tissue O_2 concentration in eq. (1.1-1).

- A spatial grid for eq. (1.1-1) is defined with 21 points so that z = 0,1/20=0.05,...,1. The BBB length is a normalized value, $z = z_u = 1$.

```
#
# Spatial grid in z
  zl=0;zu=1;dz=(zu-zl)/(nz-1);
  z=seq(from=zl,to=zu,by=dz);
```

- An interval in t is defined for 21 output points so that tout=0,1/20=0.05, ...,1. The time scale is normalized with $t_f = 1$ specified as the final time that is considered appropriate, e.g., day, week, and month.

```
#
# Independent variable for ODE integration
  t0=0;tf=1;nout=21;
  tout=seq(from=t0,to=tf,by=(tf-t0)/(nout-1));
```

- IC (1.1-2) is implemented, with u_{1n} = u1n defined previously as a parameter.

```
#
# Initial condition (t=0)
  u0=rep(u1n,nz);
  ncall=0;
```

Also, the counter for the calls to pde1a is initialized.

- The system of nz=21 ODEs is integrated by the library integrator lsodes (available in deSolve, [2]). As expected, the inputs to lsodes are the ODE function, pde1a, the IC vector u0, and the vector of output values of t, tout. The length of u0 (21) informs lsodes how many ODEs are to be integrated. func,y,times are reserved names.

Implementation of the One PDE Model

```
#
# ODE integration
  out=lsodes(y=u0,times=tout,func=pde1a,
       sparsetype="sparseint",rtol=1e-6,
       atol=1e-6,maxord=5);
  nrow(out)
  ncol(out)
```

- nrow,ncol confirm the dimensions of out.
- $u_1(z,t)$ is placed in a matrix for subsequent plotting.

```
#
# Arrays for plotting numerical solution
  u1=matrix(0,nrow=nz,ncol=nout);
  for(it in 1:nout){
    for(iz in 1:nz){
      u1[iz,it]=out[it,iz+1];
    }
    u1[1,it]=u1e;
  }
```

The offset +1 is required because the first element of the solution vectors in out is the value of t and the 2 to 22 elements are the 21 values of u_1. These dimensions from the preceding calls to nrow,ncol are confirmed in the subsequent output.
- The numerical values of $u_1(z,t)$ returned by lsodes are displayed. Every fourth value in t and every second value in z appear from by=4,2.

```
#
# Display numerical solution
  iv=seq(from=1,to=nout,by=4);
  for(it in iv){
    cat(sprintf("\n      t       z    u1(z,t)\n"));
    iv=seq(from=1,to=nz,by=2);
    for(iz in iv){
      cat(sprintf("%6.2f%6.2f%12.3e\n",
          tout[it],z[iz],u1[iz,it]));
    }
  }
```

- The number of calls to pde1a is displayed at the end of the solution.

```
#
# Calls to ODE routine
  cat(sprintf("\n\n ncall = %5d\n\n",ncall));
```

- $u_1(z,t)$ is plotted in 2D against z and parametrically in t with the R utility `matplot` and in 3D with R utility `persp`. `par(mfrow=c(1,1))` specifies a 1×1 matrix of plots, that is, one plot on a page.

```
#
# Plot PDE solution
#
# u1
  par(mfrow=c(1,1));
  matplot(x=z[2:nz],y=u1[2:nz,],type="l",xlab="z",
          ylab="u1(z,t)",xlim=c(zl,zu),lty=1,main="",lwd=2,
          col="black");
  persp(z,tout,u1,theta=60,phi=45,
        xlim=c(zl,zu),ylim=c(t0,tf),zlim=c(0,1.1),
        xlab="z",ylab="t",zlab="u1(z,t)");
```

This completes the discussion of the main program for eqs. (1.1). The ODE/MOL routine `pde1a` called by `lsodes` from the main program for the numerical MOL integration of eqs. (1.1) is next.

ODE/MOL routine

`pde1a` called in the main program of Listing 2.1 follows.

```
  pde1a=function(t,u,parm){
#
# Function pde1a computes the t derivative
# of u1(z,t)
#
# One vector to one vector
  u1=rep(0,nz);
  for(iz in 1:nz){
    u1[iz]=u[iz];}
#
# BC, z=zl
  u1[1]=u1e;
#
# PDE
  u1t=rep(0,nz);
  for(iz in 1:nz){
    if(iz==1){u1t[iz]=0;}
    if(iz>1){
      u1t[iz]=-vz*(u1[iz]-u1[iz-1])/dz-
              (2/r1)*km1*(u1[iz]-u2n);}
  }
```

Implementation of the One PDE Model

```
#
# One vector to one vector
  ut=rep(0,nz);
  for(iz in 1:nz){
    ut[iz]=u1t[iz];}
#
# Increment calls to pde1a
  ncall <<- ncall+1;
#
# Return derivative vector
  return(list(c(ut)));
  }
```

Listing 2.2: ODE/MOL routine for eqs. (1.1)

We can note the following details about Listing 2.2.

- The function is defined.

    ```
    pde1a=function(t,u,parm){
    #
    # Function pde1a computes the t derivative
    # of u1(z,t)
    ```

    t is the current value of t in eqs. (1.1). u is the 21-vector of ODE/PDE dependent variables. parm is an argument to pass parameters to pde1a (unused, but required in the argument list). The arguments must be listed in the order stated to properly interface with lsodes called in the main program of Listing 2.1. The derivative vector of the LHS of eq. (1.1-1) is calculated and returned to lsodes as explained subsequently.

- BC (1.1-3) is programmed.

    ```
    #
    # BC, z=zl
      u1[1]=u1e;
    ```

- Equation (1.1-1) is programmed.

    ```
    #
    # PDE
      u1t=rep(0,nz);
      for(iz in 1:nz){
        if(iz==1){u1t[iz]=0;}
        if(iz>1){
          u1t[iz]=-vz*(u1[iz]-u1[iz-1])/dz-
                  (2/r1)*km1*(u1[iz]-u2n);}
      }
    ```

The derivative $\frac{\partial u_1}{\partial z}$ in eq. (1.1-1) is approximated with a two-point upwind finite difference (FD).

$$\frac{\partial u_1(z,t)}{\partial z} \approx \frac{u_1(z,t) - u_1(z-\Delta z,t)}{\Delta z} + O(\Delta z)$$

programmed as

(u1[iz]-u1[iz-1])/dz

$O(\Delta z)$ indicates that the error in the FD approximation is first order in Δz. The variation in the numerical solution of eq. (1.1-1) can be studied as a function of the FD increment Δz by varying nz in Listing 2.1. Alternate approximations for the axial derivative $\frac{\partial u_1}{\partial z}$ are considered in Appendix A and [1].

For $z = z_l$, BC (1.1-3) sets the value of $u_1(z = z_l, t)$ and therefore the derivative is set to zero (if(iz==1)u1t[iz]=0;) to ensure the boundary value is maintained.

The correspondence of the programming with eq. (1.1-1) is an important feature of the MOL.

- The 21 ODE derivatives are placed in the vector ut for return to lsodes to take the next in t along the solution.

```
#
# One vector to one vector
  ut=rep(0,nz);
  for(iz in 1:nz){
    ut[iz]=u1t[iz];}
```

- The counter for the calls to pde1a is incremented and returned to the main program of Listing 2.1 by <<-.

```
#
# Increment calls to pde1a
  ncall <<- ncall+1;
```

- The vector ut is returned as a list as required by lsodes. c is the R vector utility. The final } concludes pde1a.

```
#
# Return derivative vector
  return(list(c(ut)));
  }
```

This completes the discussion of pde1a. The output from the main program of Listing 2.1 and ODE/MOL routine pde1a of Listing 2.2 is considered next.

Numerical, graphical output

The numerical output is given in Table 2.1.

[1] 21

[1] 22

t	z	u1(z,t)
0.00	0.00	7.500e-01
0.00	0.10	1.000e+00
0.00	0.20	1.000e+00
0.00	0.30	1.000e+00
0.00	0.40	1.000e+00
0.00	0.50	1.000e+00
0.00	0.60	1.000e+00
0.00	0.70	1.000e+00
0.00	0.80	1.000e+00
0.00	0.90	1.000e+00
0.00	1.00	1.000e+00

t	z	u1(z,t)
0.20	0.00	7.500e-01
0.20	0.10	6.401e-01
0.20	0.20	6.187e-01
0.20	0.30	6.449e-01
0.20	0.40	6.635e-01
0.20	0.50	6.692e-01
0.20	0.60	6.702e-01
0.20	0.70	6.703e-01
0.20	0.80	6.703e-01
0.20	0.90	6.703e-01
0.20	1.00	6.703e-01

. .
. .
. .

Output for t = 0.4, 0.6,
        0.8 removed

. .
. .
. .

t	z	u1(z,t)
1.00	0.00	7.500e-01
1.00	0.10	6.198e-01
1.00	0.20	5.123e-01

```
  1.00    0.30    4.234e-01
  1.00    0.40    3.499e-01
  1.00    0.50    2.894e-01
  1.00    0.60    2.400e-01
  1.00    0.70    2.010e-01
  1.00    0.80    1.719e-01
  1.00    0.90    1.523e-01
  1.00    1.00    1.410e-01

ncall =     113
```

Table 2.1: Numerical output from Listings 2.1, 2.2

We can note the following details about this output.

- 21 t output points as the first dimension of the solution matrix out from lsodes as programmed in the main program of Listing 2.1 (with nout=21).
- The solution matrix out returned by lsodes has 22 elements as a second dimension. The first element is the value of t. Elements 2 to 22 are $u_1(z,t)$ from eqs. (1.1) (for each of the 21 output points).
- The solution is displayed for t=0,1/20=0.05,...,1 as programmed in Listing 2.1 (every fourth value of t is displayed as explained previously).
- The solution is displayed for z=0,1/20=0.05,...,1 as programmed in Listing 2.1 (every second value of z is displayed as explained previously).
- IC (1.1-2) is confirmed ($t = 0$).
- BC (1.1-3) is confirmed ($t = 0, z = z_l = 0$). There is a discontinuous change from the IC ($t = 0$) to the BC ($t > 0$), which is subsequently smoothed, e.g., $t = 0.2$. To improve the appearance of the graphical output (Figure 2.1-1 considered next), the output does not include the solution at $z = z_l = 0$ (matplot(x=z[2:nz],y=u1[2:nz,],... in Listing 2.1). Use of Listings 2.1, 2.2 with the solution at $z = z_l = 0$ included is left as an exercise.
- The computational effort as indicated by ncall = 113 is modest so that lsodes computed the solution to eqs. (1.1) efficiently.

The graphical output is in Figures 2.1.

The solution $u_1(z,t)$ starts from IC (1.1-2), and confirms BC (1.1-3). In general, $u_1(z,t)$ is a front moving left to right (as expected with $v_z > 0$) in response to the reduced entering O_2 concentration, $u_1(z = z_l = 0, t) = u_{1e} = 0.75$.

Figure 2.1-2 reflects the decreasing blood O_2 concentration with t.

Two special cases can be considered to confirm the model and coding:

- $v_z = 0$: With no blood flood, $u_1(z,t)$ remains at IC (1.1-2).
- $k_{m1} = 0$: With no O_2 transfer through the BBB, $u_1(z, t \to \infty) = u_{1e}$ ($t_f = 2$) indicates the solution for large t.

These cases are left as exercises.

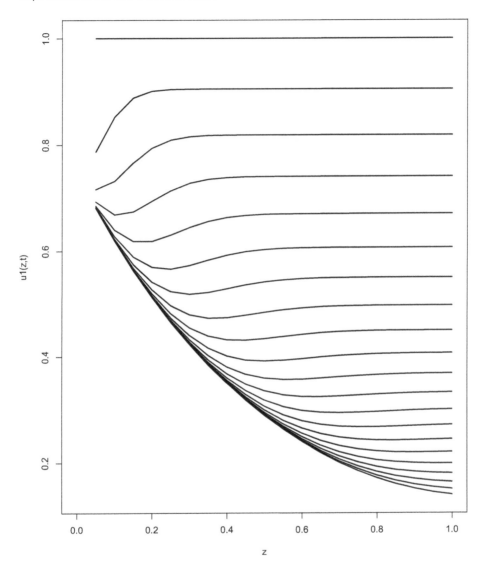

Figure 2.1-1 $u_1(z,t)$ from eqs. (1.1), 2D.

2.1.2 SUMMARY AND CONCLUSIONS

The one PDE model for blood O_2 concentration, $u_1(z,t)$ defined by eqs. (1.1), is implemented within the MOL. Parameters in this model are defined numerically in Listing 2.1 and can be used to investigate dynamic effects, e.g., the blood convection with v_z, mass transfer through the BBB with k_{m1}.

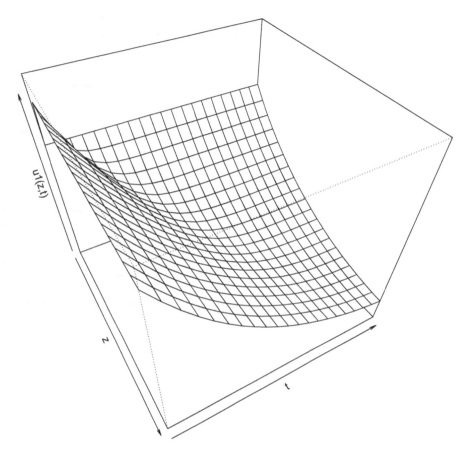

Figure 2.1-2 $u_1(z,t)$ from eqs. (1.1), 3D.

For this prototype code, which is intended to explain the computer implementation with the main program of Listing 2.1 and ODE/MOL routine of Listing 2.2, the brain tissue O_2 concentration is constant ($u_{2n} = 0$ in eq. (1.1-1)). The model is now extended to include a second PDE for the brain tissue O_2 concentration, $u_2(z,t)$.

REFERENCES

1. Schiesser, W.E. (2019), *Numerical PDE Analysis of the Blood Brain Barrier*, World Scientific Publishing Co., Singapore.
2. Soetaert, K., J. Cash, and F. Mazzia (2012), *Solving Differential Equations in R*, Springer-Verlag, Heidelberg, Germany.

3 Two PDE Model

3.1 INTRODUCTION

The one PDE model of Chapter 1, eqs. (1.1), is extended in this chapter by adding a PDE for the brain tissue O_2 concentration, $u_2(z,t)$. Equation (1.1-1) for $u_1(z,t)$ is also modified to include $u_2(z,t)$.

3.1.1 TWO PDE MODEL FORMULATION

Equations (1.1) are restated with $u_2(z,t)$ included.

$$\frac{\partial u_1}{\partial t} = -v_z \frac{\partial u_1}{\partial z} - (2/r_l)k_{m1}(u_1 - u_2) \qquad (3.1\text{-}1)$$

$$u_1(z, t = 0) = u_{1n} \qquad (3.1\text{-}2)$$

$$u_1(z = z_l, t) = u_{1e}(t) \qquad (3.1\text{-}3)$$

A mass balance on the O_2 concentration in the brain tissue follows (the balance is derived in chapter appendix A3):

$$\frac{\partial u_2}{\partial t} = D_2 \frac{\partial^2 u_2}{\partial z^2} + \frac{2r_l}{(r_u^2 - r_l^2)} k_{m1}(u_1 - u_2) \qquad (3.2\text{-}1)$$

Equation (3.2-1) states that the time rate of change of the brain tissue O_2 concentration $\left(\dfrac{\partial u_2}{\partial t}\right)$ equals the sum of the rate of axial diffusion of O_2 along the brain tissue $\left(D_2 \dfrac{\partial^2 u_2}{\partial z^2}\right)$ and the rate of transfer of O_2 across the BBB $\left(\dfrac{2r_l}{(r_u^2 - r_l^2)} k_{m1}(u_1 - u_2)\right)$. D_2 is the effective diffusivity (dispersion coefficient) for O_2 in the axial (longitudinal z) direction. Radial (transverse to z) variation of O_2 has been neglected, which implies a narrow radial dimension so that the O_2 concentration is effectively uniform in the radial direction[1].

Equation (3.2-1) is first order in t so that it requires one initial condition (IC).

$$u_2(z, t = 0) = u_{2n} \qquad (3.2\text{-}2)$$

u_{2n} is the normalized brain tissue O_2 concentration used as an IC.

Equation (3.2-1) is second order in z and requires two boundary conditions (BCs).

$$\frac{\partial u_2(z = z_l, t)}{\partial z} = \frac{\partial u_2(z = z_u, t)}{\partial z} = 0 \qquad (3.2\text{-}3,4)$$

---

[1] Radial variation of O_2 in the brain tissue can be analyzed by the methodology discussed in [1].

Equations (3.2-3,4) are homogeneous (zero) Neumann BCs for zero flux at the boundaries of the brain tissue.

Equations (3.1), (3.2) constitute the two PDE model that is implemented in R, as explained in Chapter 4.

3.1.2 SUMMARY AND CONCLUSIONS

Equations (3.1), (3.2) are the two PDE model for O_2 in the capillary blood flow and brain tissue. The transfer of O_2 across the BBB is represented with the terms $\pm k_{m1}(u_1 - u_2)$ which can be used to study the effects of reduced O_2 from the blood stream to the brain tissue resulting from pulmonary impairment (through variation of k_{m1}).

The implementation of eqs. (3.1), (3.2) is considered in Chapter 4.

APPENDIX A3: DERIVATION OF THE BRAIN TISSUE O_2 BALANCE

A mass balance on the brain tissue with the incremental volume $(\pi r_u^2 - \pi r_l^2)\Delta z$ (r_l, r_u are the inner and outer radial boundaries of the brain tissue, respectively) gives

$$(\pi r_u^2 - \pi r_l^2)\Delta z \frac{\partial u_2(z,t)}{\partial t} = (\pi r_u^2 - \pi r_l^2)\left(-D_2 \frac{\partial u_2(z,t)}{\partial z}\bigg|_z + D_2 \frac{\partial u_2(z,t)}{\partial z}\bigg|_{z+\Delta z}\right)$$
$$+ 2\pi r_l \Delta z k_{m1}(u_1(z,t) - u_2(z,t)) \quad \text{(A3.1)}$$

The terms in eq. (A3.1) are

- $(\pi r_u^2 - \pi r_l^2)\Delta z \frac{\partial u_2(z,t)}{\partial t}$: accumulation of O_2 in the incremental volume $(\pi r_u^2 - \pi r_l^2)\Delta z$. If this term is negative (from the sum of the RHS terms), the O_2 concentration decreases (is depleted) with time.
- $(\pi r_u^2 - \pi r_l^2)\left(-D_2 \frac{\partial u_2(z,t)}{\partial z}\bigg|_z\right)$: rate of diffusion of O_2 in the brain tissue at z. This term is based on Fick's first law

$$q = -D_2 \frac{\partial u_2(z,t)}{\partial z} \quad \text{(A3.2)}$$

- $(\pi r_u^2 - \pi r_l^2)\left(D_2 \frac{\partial u_2(z,t)}{\partial z}\bigg|_{z+\Delta z}\right)$: rate of diffusion of O_2 in the brain tissue at $z + \Delta z$.
- $2\pi r_l \Delta z k_{m1}(u_1(z,t) - u_2(z,t))$: rate of mass transfer of the O_2 transported between the capillary and the inner surface of the BBB. The transfer area is $2\pi r_l \Delta z$ and the mass transfer coefficient is k_{m1}.

If eq. (A3.1) is divided by the incremental volume $(\pi r_u^2 - \pi r_l^2)\Delta z$ (the coefficient of $\dfrac{\partial u_2(z,t)}{\partial t}$), after minor rearrangement,

$$\frac{\partial u_2(z,t)}{\partial t} = \frac{D_2 \dfrac{\partial u_2(z,t)}{\partial z}\Big|_{z+\Delta z} - D_2 \dfrac{\partial u_2(z,t)}{\partial z}\Big|_z}{\Delta z} + \frac{2r_l}{(r_u^2 - r_l^2)} k_{m1}(u_1(z,t) - u_2(z,t)) \qquad (A3.3)$$

For $\Delta z \to 0$, eq. (A3.3) becomes

$$\frac{\partial u_2(z,t)}{\partial t} = D_2 \frac{\partial^2 u_2(z,t)}{\partial z^2} + \frac{2r_l}{(r_u^2 - r_l^2)} k_{m1}(u_1(z,t) - u_2(z,t)) \qquad (A3.4)$$

Equation (A3.4) is eq. (3.2-1).

REFERENCE

1. Schiesser, W.E. (2019), *Numerical PDE Analysis of the Blood Brain Barrier*, World Scientific Publishing Co., Singapore.

4 Implementation of the Two PDE Model

4.1 INTRODUCTION

The two PDE model of Chapter 3 for O_2 in the capillary blood (eqs. (3.1)) and the brain tissue (eqs. (3.2)) is implemented in R routines as discussed in this chapter.

4.1.1 R ROUTINES FOR THE TWO PDE MODEL

Equations (3.1), (3.2) for the O_2 concentration in the capillary blood and brain tissue are implemented with the following R routines, starting with a main program.

Main program

The main program for eqs. (3.1), (3.2) follows:

```
#
#  Two PDE model
#
# Delete previous workspaces
  rm(list=ls(all=TRUE))
#
# Access ODE integrator
  library("deSolve");
#
# Access functions for numerical solution
  setwd("f:/brain hypoxia/chap4");
  source("pde2a.R");
#
# Parameters
  nz=21;
  rl=1;
  ru=2;
  D2=0.1;
  km1=0;
  km1=10;
  vz=1;
  u1e=0.75;
  u1n=1;
  u2n=1;
#
```

```
# Constants
  r1=2/rl;
  r2=(2*rl)/(ru^2-rl^2);
#
# Spatial grid in z
  zl=0;zu=1;dz=(zu-zl)/(nz-1);dzs=dz^2;
  z=seq(from=zl,to=zu,by=dz);
#
# Independent variable for ODE integration
  t0=0;tf=1;nout=21;
  tout=seq(from=t0,to=tf,by=(tf-t0)/(nout-1));
#
# Initial condition (t=0)
  u0=rep(0,2*nz);
  for(iz in 1:nz){
    u0[iz]    =u1n;
    u0[iz+nz]=u2n;
  }
  ncall=0;
#
# ODE integration
  out=lsodes(y=u0,times=tout,func=pde2a,
      sparsetype="sparseint",rtol=1e-6,
      atol=1e-6,maxord=5);
  nrow(out)
  ncol(out)
#
# Arrays for plotting numerical solution
  u1=matrix(0,nrow=nz,ncol=nout);
  u2=matrix(0,nrow=nz,ncol=nout);
  for(it in 1:nout){
    for(iz in 1:nz){
      u1[iz,it]=out[it,iz+1];
      u2[iz,it]=out[it,iz+1+nz];
    }
   u1[1,it]=u1e;
  }
#
# Display numerical solution
  iv=seq(from=1,to=nout,by=4);
  for(it in iv){
    cat(sprintf("\n    t      z      u1(z,t)      u2(z,t)\n"));
    iv=seq(from=1,to=nz,by=2);
    for(iz in iv){
```

```
      cat(sprintf("%6.2f %6.2f %12.3e %12.3e\n",
          tout[it],z[iz],u1[iz,it],u2[iz,it]));
    }
  }
#
# Calls to ODE routine
  cat(sprintf("\n\n ncall = %5d\n\n",ncall));
#
# Plot PDE solution
#
# u1
  par(mfrow=c(1,1));
  matplot(x=z[2:nz],y=u1[2:nz,],type="l",xlab="z",ylab="u1(z,t)",
          xlim=c(zl,zu),lty=1,main="",lwd=2,col="black");
  persp(z,tout,u1,theta=60,phi=45,
        xlim=c(zl,zu),ylim=c(t0,tf),zlim=c(0,1.1),
        xlab="z",ylab="t",zlab="u1(z,t)");
#
# u2
  par(mfrow=c(1,1));
  matplot(x=z,y=u2,type="l",xlab="z",ylab="u2(z,t)",
          xlim=c(zl,zu),lty=1,main="",lwd=2,col="black");
  persp(z,tout,u2,theta=60,phi=45,
        xlim=c(zl,zu),ylim=c(t0,tf),zlim=c(0,1.1),
        xlab="z",ylab="t",zlab="u2(z,t)");
```

Listing 4.1: Main program for eqs. (3.1), (3.2)

We can note the following details about Listing 4.1 (with some repetition of the discussion of Listing 2.1 so that the explanation is self-contained).

- Previous workspaces are deleted.

    ```
    #
    #  Two PDE model
    #
    # Delete previous workspaces
      rm(list=ls(all=TRUE))
    ```

- The R ODE integrator library deSolve is accessed [3].

    ```
    #
    # Access ODE integrator
      library("deSolve");
    #
    # Access functions for numerical solution
    ```

```
setwd("f:/brain hypoxia/chap4");
source("pde2a.R");
```

Then the directory with the files for the solution of eqs. (3.1), (3.2) is designated. Note that setwd (set working directory) uses / rather than the usual \.

- The model parameters are specified numerically.

```
#
# Parameters
  nz=21;
  rl=1;
  ru=2;
  D2=0.1;
  km1=0;
  km1=10;
  vz=1;
  u1e=0.75;
  u1n=1;
  u2n=1;
```

where

- nz: number of spatial grid points for eqs. (3.1-1), (3.2-1).
- rl: radius of the blood capillary and inner radius of the blood brain barrier (BBB). The wall thickness of the BBB is neglected so r_l = rl also designates the inner boundary of the brain tissue.
- ru: outer boundary of the brain tissue, r_u = ru.
- D_2: O_2 diffusivity of the brain tissue in eq. (3.2-1).
- km1: mass transfer (permeability) coefficient for O_2 across the BBB in eqs. (3.1-1), (3.2-1). For $k_m = 0$, O_2 is not transferred from the capillary blood to the brain tissue, so the latter remains at the IC of eq. (3.2-2), $u_2(z,t=0) = u_{2n}$.

  To further explain this special case, BCs (3.2-3,4) are consistent with the constant solution in z, $u_2(z,t=0) = u_{2n} = u_2(z,t>0)$. Also, for a constant solution in z, $\frac{\partial^2 u_2}{\partial z^2} = 0$ in eq. (3.2-1). Thus, the RHS terms of eq. (3.2-1) are zero (with $k_{m1} = 0$) so $\frac{\partial u_2}{\partial t} = 0$ (from eq. (3.2-1)), and the solution remains at IC eq. (3.2-2).

  This special case is an important check since a departure of $u_2(z,t)$ from IC would indicate a coding error (in pde2a of Listing 4.2 that follows). This special case is left as an exercise.
- vz: capillary blood flow superficial velocity in eq. (3.1-1).
- u1e: entering blood O_2 concentration in BC (3.1-3).
- u1n: normalized blood O_2 concentration in IC (3.1-2).
- u2n: normalized brain tissue O_2 concentration in IC (3.2-2).

- The constants $r_1 = \dfrac{2}{r_l}$, $r_2 = \dfrac{2r_l}{(r_u^2 - r_l^2)}$ are computed for use in the ordinary differential equation/method of lines (ODE/MOL) routine pde2a.

```
#
# Constants
  r1=2/rl;
  r2=(2*rl)/(ru^2-rl^2);
```

- A spatial grid for eqs. (3.1-1), (3.2-1) is defined with 21 points so that z = 0,1/20=0.05,...,1. The BBB length is a normalized value, $z = z_u = 1$.

```
#
# Spatial grid in z
  zl=0;zu=1;dz=(zu-zl)/(nz-1);dzs=dz^2;
  z=seq(from=zl,to=zu,by=dz);
```

- An interval in t is defined for 21 output points so that tout=0,1/20=0.05, ...,1. The time scale is normalized with $t_f = 1$ specified as the final time that is considered appropriate, e.g., day, week, or month.

```
#
# Independent variable for ODE integration
  t0=0;tf=1;nout=21;
  tout=seq(from=t0,to=tf,by=(tf-t0)/(nout-1));
```

- ICs (3.1-2), (3.2-2) are implemented (u1n,u2n are defined previously as parameters).

```
#
# Initial condition (t=0)
  u0=rep(0,2*nz);
  for(iz in 1:nz){
    u0[iz]    =u1n;
    u0[iz+nz] =u2n;
  }
  ncall=0;
```

Also, the counter for the calls to ode2a is initialized.
- The system of 2*nz=42 ODEs is integrated by the library integrator lsodes (available in deSolve, [3]). As expected, the inputs to lsodes are the ODE function, pde2a, the IC vector u0, and the vector of output values of t, tout. The length of u0 (42) informs lsodes how many ODEs are to be integrated. func,y,times are reserved names.

```
#
# ODE integration
  out=lsodes(y=u0,times=tout,func=pde2a,
      sparsetype="sparseint",rtol=1e-6,
      atol=1e-6,maxord=5);
  nrow(out)
  ncol(out)
```

`nrow,ncol` confirm the dimensions of `out`.

- $u_1(z,t)$, $u_2(z,t)$ are placed in matrices for subsequent plotting.

```
#
# Arrays for plotting numerical solution
  u1=matrix(0,nrow=nz,ncol=nout);
  u2=matrix(0,nrow=nz,ncol=nout);
  for(it in 1:nout){
    for(iz in 1:nz){
      u1[iz,it]=out[it,iz+1];
      u2[iz,it]=out[it,iz+1+nz];
    }
    u1[1,it]=u1e;
  }
```

The offset +1 is required because the first element of the solution vectors in `out` is the value of t and the 2 to 43 elements are the 42 values of u_1, u_2. These dimensions from the preceding calls to `nrow,ncol` are confirmed in the subsequent output.

- The numerical values of $u_1(z,t)$, $u_2(z,t)$ returned by `lsodes` are displayed. Every fourth value in t and every second value in z appear from `by=4,2`.

```
#
# Display numerical solution
  iv=seq(from=1,to=nout,by=4);
  for(it in iv){
    cat(sprintf("\n      t       z      u1(z,t)     u2(z,t)\n"));
    iv=seq(from=1,to=nz,by=2);
    for(iz in iv){
      cat(sprintf("%6.2f %6.2f %12.3e %12.3e\n",
          tout[it],z[iz],u1[iz,it],u2[iz,it]));
    }
  }
```

- The number of calls to `pde2a` is displayed at the end of the solution.

```
#
# Calls to ODE routine
  cat(sprintf("\n\n ncall = %5d\n\n",ncall));
```

Implementation of the Two PDE Model

- $u_1(z,t)$, $u_2(z,t)$ are plotted in 2D against z and parametrically in t with the R utility `matplot` and in 3D with R utility `persp`. `par(mfrow=c(1,1))` specifies a 1×1 matrix of plots, that is, one plot on a page.

```
#
# Plot PDE solutions
#
# u1
  par(mfrow=c(1,1));
  matplot(x=z[2:nz],y=u1[2:nz,],type="l",xlab="z",
          ylab="u1(z,t)",xlim=c(zl,zu),lty=1,main="",lwd=2,
          col="black");
  persp(z,tout,u1,theta=60,phi=45,
        xlim=c(zl,zu),ylim=c(t0,tf),zlim=c(0,1.1),
        xlab="z",ylab="t",zlab="u1(z,t)");
#
# u2
  par(mfrow=c(1,1));
  matplot(x=z,y=u2,type="l",xlab="z",ylab="u2(z,t)",
          xlim=c(zl,zu),lty=1,main="",lwd=2,col="black");
  persp(z,tout,u2,theta=60,phi=45,
        xlim=c(zl,zu),ylim=c(t0,tf),zlim=c(0,1.1),
        xlab="z",ylab="t",zlab="u2(z,t)");
```

The 2D plot of $u_1(z,t)$ does not include the discontinuous change from IC (3.1-2) to BC (3.1-3), `x=z[2:nz]`,`y=u1[2:nz,]`. This discontinuity is not included in $u_2(z,t)$, so the solution at $z = z_l$ is included, `x=z,y=u2`.

This completes the discussion of the main program for eqs. (3.1), (3.2). The ODE/MOL routine pde2a called by lsodes from the main program for the numerical MOL integration of eqs. (3.1), (3.2) is discussed in the following section.

ODE/MOL routine

pde2a called in the main program of Listing 4.1 follows:

```
  pde2a=function(t,u,parm){
#
# Function pde2a computes the t derivatives
# of u1(z,t),u2(z,t)
#
# One vector to two vectors
  u1=rep(0,nz);
  u2=rep(0,nz);
  for(iz in 1:nz){
    u1[iz]=u[iz];
```

```
    u2[iz]=u[iz+nz];
  }
#
# BC, z=zl
  u1[1]=u1e;
#
# PDEs
  u1t=rep(0,nz);
  for(iz in 1:nz){
    if(iz==1){u1t[1]=0;}
    if(iz>1){
      u1t[iz]=-vz*(u1[iz]-u1[iz-1])/dz-
        r1*km1*(u1[iz]-u2[iz]);}
  }
  u2t=rep(0,nz);
  for(iz in 1:nz){
    if(iz==1){
      u2t[1]=
        2*D2*(u2[2]-u2[1])/dzs+
        r2*km1*(u1[1]-u2[1]);}
    if(iz==nz){
      u2t[nz]=
        D2*2*(u2[nz-1]-u2[nz])/dzs+
        r2*km1*(u1[nz]-u2[nz]);}
    if((iz>1)&&(iz<nz)){
      u2t[iz]=
        D2*(u2[iz+1]-2*u2[iz]+u2[iz-1])/dzs+
        r2*km1*(u1[iz]-u2[iz]);}
  }
#
# Two vectors to one vector
  ut=rep(0,2*nz);
  for(iz in 1:nz){
    ut[iz]   =u1t[iz];
    ut[iz+nz]=u2t[iz];
  }
#
# Increment calls to pde2a
  ncall <<- ncall+1;
#
# Return derivative vector
  return(list(c(ut)));
  }
```

Listing 4.2: ODE/MOL routine for eqs. (3.1), (3.2)

Implementation of the Two PDE Model

We can note the following details about Listing 4.2.

- The function is defined.

  ```
  pde2a=function(t,u,parm){
  #
  # Function pde2a computes the t derivatives
  # of u1(z,t),u2(z,t)
  ```

  t is the current value of t in eqs. (3.1), (3.2). u is the 42-vector of ODE/PDE dependent variables. parm is an argument to pass parameters to pde2a (unused, but required in the argument list). The arguments must be listed in the order stated to properly interface with lsodes called in the main program of Listing 4.1. The derivative vector of the LHS of eqs. (3.1-1), (3.2-1) is calculated and returned to lsodes as explained subsequently.

- u is placed in two vectors to facilitate the programming of eqs. (3.1), (3.2).

  ```
  #
  # One vector to two vectors
    u1=rep(0,nz);
    u2=rep(0,nz);
    for(iz in 1:nz){
      u1[iz]=u[iz];
      u2[iz]=u[iz+nz];
    }
  ```

- BC (3.1-3) is programmed.

  ```
  #
  # BC, z=zl
    u1[1]=u1e;
  ```

- Equation (3.1-1) is programmed.

  ```
  #
  # PDEs
    u1t=rep(0,nz);
    for(iz in 1:nz){
      if(iz==1){u1t[1]=0;}
      if(iz>1){
        u1t[iz]=-vz*(u1[iz]-u1[iz-1])/dz-
          r1*km1*(u1[iz]-u2[iz]);}
    }
  ```

  The derivative $\dfrac{\partial u_1}{\partial z}$ in eq. (3.1-1) is approximated with a two point upwind finite difference (FD).

  $$\frac{\partial u_1(z,t)}{\partial z} \approx \frac{u_1(z,t)-u_1(z-\Delta z,t)}{\Delta z}+O(\Delta z)$$

programmed as

(u1[iz]-u1[iz-1])/dz

$O(\Delta z)$ indicates that the error in the FD approximation is first order in Δz. The variation in the numerical solution of eq. (3.1-1) can be studied as a function of the FD increment Δz by varying nz in Listing 4.1. Alternate approximations for the axial derivative $\frac{\partial u_1}{\partial z}$ are considered in [2]. For $z = z_l$, BC (3.1-3) sets the value of $u_1(z = z_l, t)$, and therefore, the derivative is set to zero (if(iz==1)u1t[1]=0;) to ensure the boundary value is maintained.

- Equation (3.2-1) is programmed.

```
u2t=rep(0,nz);
for(iz in 1:nz){
  if(iz==1){
    u2t[1]=
      2*D2*(u2[2]-u2[1])/dzs+
      r2*km1*(u1[1]-u2[1]);}
  if(iz==nz){
    u2t[nz]=
      D2*2*(u2[nz-1]-u2[nz])/dzs+
      r2*km1*(u1[nz]-u2[nz]);}
  if((iz>1)&&(iz<nz)){
    u2t[iz]=
      D2*(u2[iz+1]-2*u2[iz]+u2[iz-1])/dzs+
      r2*km1*(u1[iz]-u2[iz]);}
}
```

This code requires some additional explanation.

- The derivative $\frac{\partial^2 u_2}{\partial z^2}$ in eq. (3.2-1) is approximated with a three-point centered finite difference (FD).

$$\frac{\partial^2 u_2(z,t)}{\partial z^2} \approx \frac{u_2(z+\Delta z,t) - 2u_2(z,t) + u_2(z-\Delta z,t)}{\Delta z^2} + O(\Delta z^2)$$
(4.1-1)

$O(\Delta z^2)$ indicates that the error in the FD approximation is second order in Δz[1].

- BC (3.2-3) is approximated with a two-point centered FD.

$$\frac{\partial u_2(z=z_l,t)}{\partial z} \approx \frac{u_2(z_l+\Delta z,t) - u_2(z_l-\Delta z,t)}{2\Delta z} = 0$$

---

[1] Higher-order FD approximations, $O(\Delta z^p)$, $p = 4, 6, 8, 10$, are considered in [1]. As a point of terminology, an error analysis of a numerical solution based on the variation of the FD interval, $\Delta z = h$ is termed *h refinement*. Variation of the order of the FD approximation, $O(h^p)$, is termed *p refinement*.

or
$$u_2(z_l - \Delta z, t) = u_2(z_l + \Delta z, t) \quad (4.1\text{-}2)$$

- Combination of eqs. (4.1-1,2) gives the FD approximation

$$\frac{\partial^2 u_2(z=z_l,t)}{\partial z^2} \approx \frac{u_2(z_l + \Delta z, t) - 2u_2(z_l, t) + u_2(z_l - \Delta z, t)}{\Delta z^2}$$

$$= 2\frac{u_2(z_l + \Delta z, t) - u_2(z_l, t)}{\Delta z^2} \quad (4.1\text{-}3)$$

which is programmed as

```
2*(u2[2]-u2[1])/dzs
```

in
```
if(iz==1){
    u2t[1]=
        2*D2*(u2[2]-u2[1])/dzs+
        r2*km1*(u1[1]-u2[1]);}
```

- Similarly, BC (3.2-4) is approximated with a two point centered FD.

$$\frac{\partial u_2(z=z_u,t)}{\partial z} \approx \frac{u_2(z_u + \Delta z, t) - u_2(z_u - \Delta z, t)}{2\Delta z} = 0$$

or
$$u_2(z_u + \Delta z, t) = u_2(z_u - \Delta z, t) \quad (4.1\text{-}3)$$

- Combination of eqs. (4.1-1,3) gives the FD approximation

$$\frac{\partial^2 u_2(z=z_u,t)}{\partial z^2} \approx \frac{u_2(z_u + \Delta z, t) - 2u_2(z_u, t) + u_2(z_u - \Delta z, t)}{\Delta z^2}$$

$$= 2\frac{u_2(z_u - \Delta z, t) - u_2(z_u, t)}{\Delta z^2} \quad (4.1\text{-}4)$$

which is programmed as

```
2*(u2[nz-1]-u2[nz])/dzs
```

in
```
if(iz==nz){
    u2t[nz]=
        D2*2*(u2[nz-1]-u2[nz])/dzs+
        r2*km1*(u1[nz]-u2[nz]);}
```

- For the interior points $z_l < z < z_u$, the RHS of eq. (4.1-1) is programmed as

```
(u2[iz+1]-2*u2[iz]+u2[iz-1])/dzs
```

in
```
if((iz>1)&&(iz<nz)){
    u2t[iz]=
        D2*(u2[iz+1]-2*u2[iz]+u2[iz-1])/dzs+
        r2*km1*(u1[iz]-u2[iz]);}
```

In general, the correspondence of the programming with eqs. (3.1-1), (3.2-1) is an important feature of the MOL.
- The 42 ODE derivatives are placed in the vector ut for return to lsodes to take the next in *t* along the solution.

```
#
# Two vectors to one vector
  ut=rep(0,2*nz);
  for(iz in 1:nz){
    ut[iz]    =u1t[iz];
    ut[iz+nz]=u2t[iz];
  }
```

- The counter for the calls to pde2a is incremented and returned to the main program of Listing 4.1 by <<-.

```
#
# Increment calls to pde2a
  ncall <<- ncall+1;
```

- The vector ut is returned as a list as required by lsodes. c is the R vector utility. The final } concludes pde2a.

```
#
# Return derivative vector
  return(list(c(ut)));
  }
```

This completes the discussion of pde2a. The output from the main program of Listing 4.1 and ODE/MOL routine pde2a of Listing 4.2 is considered next.

Numerical, graphical output

The numerical output is given in Table 4.1.

[1] 21

[1] 43

t	z	u1(z,t)	u2(z,t)
0.00	0.00	7.500e-01	1.000e+00
0.00	0.10	1.000e+00	1.000e+00
0.00	0.20	1.000e+00	1.000e+00
0.00	0.30	1.000e+00	1.000e+00
0.00	0.40	1.000e+00	1.000e+00
0.00	0.50	1.000e+00	1.000e+00

```
0.00    0.60    1.000e+00    1.000e+00
0.00    0.70    1.000e+00    1.000e+00
0.00    0.80    1.000e+00    1.000e+00
0.00    0.90    1.000e+00    1.000e+00
0.00    1.00    1.000e+00    1.000e+00

  t       z      u1(z,t)      u2(z,t)
0.20    0.00    7.500e-01    9.087e-01
0.20    0.10    8.927e-01    9.357e-01
0.20    0.20    9.556e-01    9.682e-01
0.20    0.30    9.849e-01    9.880e-01
0.20    0.40    9.960e-01    9.964e-01
0.20    0.50    9.991e-01    9.991e-01
0.20    0.60    9.999e-01    9.998e-01
0.20    0.70    1.000e+00    1.000e+00
0.20    0.80    1.000e+00    1.000e+00
0.20    0.90    1.000e+00    1.000e+00
0.20    1.00    1.000e+00    1.000e+00
           .         .
           .         .
           .         .
Output for t = 0.4, 0.6,
        0.8 removed
           .         .
           .         .
           .         .
  t       z      u1(z,t)      u2(z,t)
1.00    0.00    7.500e-01    8.212e-01
1.00    0.10    8.140e-01    8.367e-01
1.00    0.20    8.494e-01    8.624e-01
1.00    0.30    8.792e-01    8.895e-01
1.00    0.40    9.063e-01    9.149e-01
1.00    0.50    9.303e-01    9.373e-01
1.00    0.60    9.504e-01    9.558e-01
1.00    0.70    9.663e-01    9.701e-01
1.00    0.80    9.779e-01    9.802e-01
1.00    0.90    9.855e-01    9.864e-01
1.00    1.00    9.892e-01    9.886e-01

ncall =    131
```

Table 4.1: Numerical output from Listings 4.1, 4.2

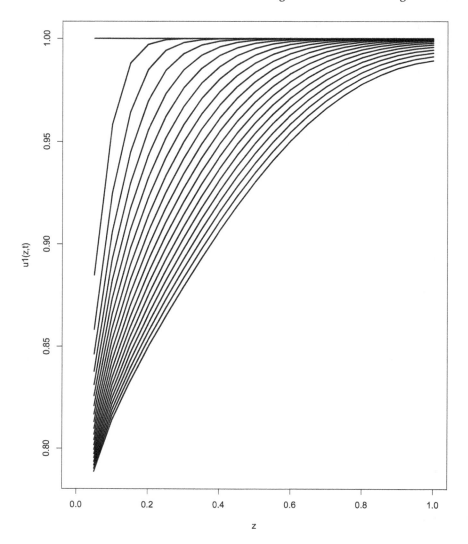

Figure 4.1-1 $u_1(z,t)$ from eqs. (3.1), 2D

We can note the following details about this output.

- 21 t output points as the first dimension of the solution matrix out from lsodes as programmed in the main program of Listing 4.1 (with nout=21).
- The solution matrix out returned by lsodes has 43 elements as a second dimension. The first element is the value of t. Elements 2 to 43 are $u_1(z,t)$, $u_2(z,t)$ from eqs. (3.1), (3.2) (for each of the 21 output points).

Implementation of the Two PDE Model

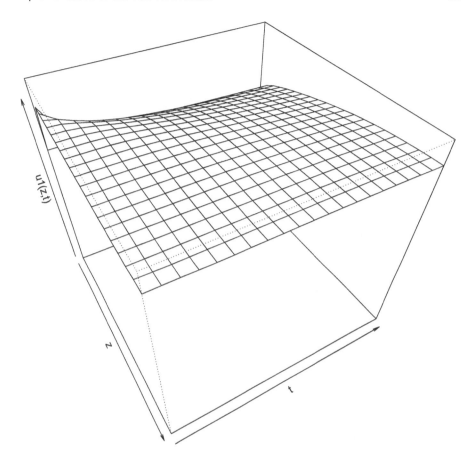

Figure 4.1-2 $u_1(z,t)$ from eqs. (3.1), 3D

- The solution is displayed for t=0,1/20=0.05,...,1 as programmed in Listing 4.1 (every fourth value of t is displayed as explained previously).
- The solution is displayed for z=0,1/20=0.05,...,1 as programmed in Listing 4.1 (every second value of z is displayed as explained previously).
- ICs (3.1-2), (3.2-2) are confirmed ($t = 0$).
- BC (3.1-3) is confirmed ($t = 0, z = z_l = 0$). There is a discontinuous change from the IC ($t = 0$) to the BC ($t > 0$), which is subsequently smoothed, e.g., $t = 0.2$. To improve the appearance of the graphical output (Figure 4.1-1 considered next), the ouput does not include the solution at $z = z_l = 0$ (matplot(x=z[2:nz],y=u1[2:nz,],... in Listing 4.1). Use of Listings 4.1, 4.2 with the solution at $z = z_l = 0$ included is left as an exercise.
- The computational effort as indicated by ncall = 131 is modest so that lsodes computed the solution to eqs. (3.1), (3.2) efficiently.

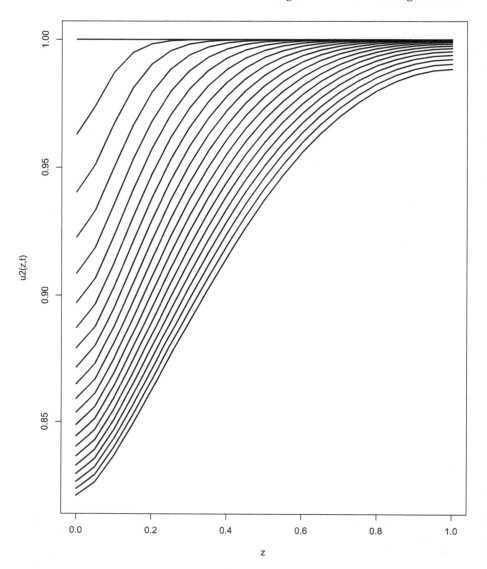

Figure 4.2-1 $u_2(z,t)$ from eqs. (3.2), 2D

The graphical output is shown in Figures 4.1, 4.2.

The solution $u_1(z,t)$ starts from IC (3.1-2) and confirms BC (3.1-3). In general, $u_1(z,t)$ is a front moving left to right (as expected with $v_z > 0$) in response to the reduced entering blood O_2 concentration, $u_1(z = z_l = 0, t) = u_{1e} = 0.75$.

Figure 4.1-2 reflects the decreasing blood O_2 concentration with t.

Implementation of the Two PDE Model

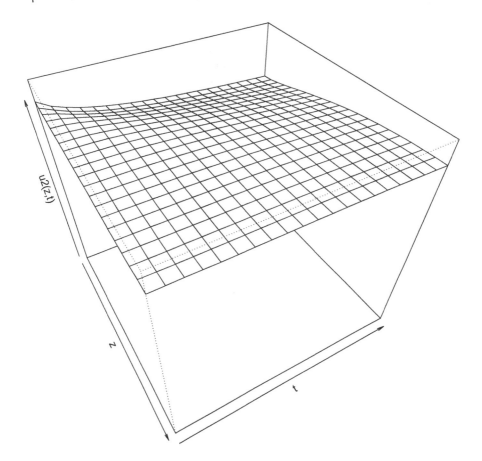

Figure 4.2-2 $u_2(z,t)$ from eqs. (3.1), 3D

The solution $u_2(z,t)$ starts from IC (3.2-2). In general, $u_2(z,t)$ is a front moving left to right in response to the reduced entering blood O_2 concentration, $u_1(z=z_l=0,t) = u_{1e} = 0.75$.

Figure 4.2-2 reflects the decreasing brain tissue O_2 concentration with t.

Two special cases can be considered to confirm the model and coding:

- $v_z = 0$: With no blood flood, $u_1(z,t)$, $u_2(z,t)$ depart from ICs (3.1-2), (3.2-2) as a result of the diffusion in z of eq. (3.2-1).
- $k_{m1} = 0$: With no O_2 transfer through the BBB, $u_1(z,t)$ is a front moving left to right ($v_z > 0$) with the long time solution $u_1(z,t \to \infty) = u_{1e}$ ($t_f = 2$ indicates the solution for large t). $u_2(z,t)$ remains at the IC (3.2-2) since there is no transfer of O_2 between the blood and the brain tissue.

These cases are left as exercises.

4.1.2 SUMMARY AND CONCLUSIONS

The two PDE model for blood and brain tissue O_2 concentrations, $u_1(z,t)$, $u_2(z,t)$ defined by eqs. (3.1), (3.2) is implemented within the MOL. Parameters in this model are defined numerically in Listing 4.1 and can be used to investigate dynamic effects, e.g., the blood convection with v_z, mass transfer through the BBB with k_{m1}.

As the next step in model development, a third PDE is added in Chapter 5 for the brain neuron cell density as it is affected by the brain tissue O_2 concentration, which could account for Covid cognitive impairment.

REFERENCES

1. Schiesser, W.E. (2016), *Method of Lines PDE Analysis in Biomedical Science and Engineering*, John Wiley, Hoboken, NJ.
2. Schiesser, W.E. (2019), *Numerical PDE Analysis of the Blood Brain Barrier*, World Scientific Publishing Co., Singapore.
3. Soetaert, K., J. Cash, and F. Mazzia (2012), *Solving Differential Equations in R*, Springer-Verlag, Heidelberg, Germany.

5 Three PDE Model

5.1 INTRODUCTION

The two PDE model of Chapter 3, eqs. (3.1), (3.2), is extended in this chapter by adding a PDE for the neuron cell density.

5.1.1 THREE PDE MODEL FORMULATION

The neuron cell density, $u_3(z,t)$, is modeled as

$$\frac{\partial u_3}{\partial t} = -k_{r3}(u_{2n} - u_2) \quad (5.1\text{-}1)$$

Equation (5.1-1) states that the functional neuron cell density decreases, $\frac{\partial u_3}{\partial t} < 0$, when the brain tissue O_2 concentration drops below the normal level, $-k_{r3}(u_{2n} - u_2) < 0$. k_{r3} is a rate constant relating the decrease in neuron cell density to the deficiency in the brain tissue O_2 concentration and is a key parameter in the three PDE model.

Equation (5.1-1) is first order in t and requires one IC.

$$u_3(z,t=0) = u_{3n} \quad (5.1\text{-}2)$$

Equation (5.1-1) is a PDE (rather than an ODE) since $u_3(z,t)$ is a function of z,t (and not just t), even though it does not explicitly have derivatives in z.

Equations (5.1) are added to eqs. (3.1), (3.2) to constitute the three PDE model for $u_1(z,t)$, $u_2(z,t)$, $u_3(z,t)$. The implementation of this model in R routines is considered next.

5.1.2 THREE PDE MODEL IMPLEMENTATION

A main program for eqs. (3.1), (3.2), (5.1) follows.

Main program

```
#
#  Three PDE model
#
# Delete previous workspaces
  rm(list=ls(all=TRUE))
#
# Access ODE integrator
  library("deSolve");
#
```

```r
# Access functions for numerical solution
  setwd("f:/brain hypoxia/chap5");
  source("pde3a.R");
#
# Parameters
  nz=21;
  rl=1;
  ru=2;
  D2=0.1;
  km1=0;
  km1=10;
  vz=1;
  u1e=0.75;
  kr3=1;
  u1n=1;
  u2n=1;
  u3n=1;
#
# Constants
  r1=2/rl;
  r2=(2*rl)/(ru^2-rl^2);
#
# Spatial grid in z
  zl=0;zu=1;dz=(zu-zl)/(nz-1);dzs=dz^2;
  z=seq(from=zl,to=zu,by=dz);
#
# Independent variable for ODE integration
  t0=0;tf=1;nout=21;
  tout=seq(from=t0,to=tf,by=(tf-t0)/(nout-1));
#
# Initial condition (t=0)
  u0=rep(0,3*nz);
  for(iz in 1:nz){
    u0[iz]      =u1n;
    u0[iz+nz]   =u2n;
    u0[iz+2*nz]=u3n;
  }
  ncall=0;
#
# ODE integration
  out=lsodes(y=u0,times=tout,func=pde3a,
      sparsetype="sparseint",rtol=1e-6,
      atol=1e-6,maxord=5);
  nrow(out)
```

```
    ncol(out)
#
# Arrays for plotting numerical solution
  u1=matrix(0,nrow=nz,ncol=nout);
  u2=matrix(0,nrow=nz,ncol=nout);
  u3=matrix(0,nrow=nz,ncol=nout);
  for(it in 1:nout){
    for(iz in 1:nz){
      u1[iz,it]=out[it,iz+1];
      u2[iz,it]=out[it,iz+1+nz];
      u3[iz,it]=out[it,iz+1+2*nz];
    }
   u1[1,it]=u1e;
  }
#
# Display numerical solution
  iv=seq(from=1,to=nout,by=4);
  for(it in iv){
    cat(sprintf("\n      t       z       u1(z,t)       u2(z,t)
                u3(z,t)\n"));
    iv=seq(from=1,to=nz,by=2);
    for(iz in iv){
      cat(sprintf("%6.2f %6.2f %12.3e %12.3e %12.3e \n",
          tout[it],z[iz],u1[iz,it],u2[iz,it],u3[iz,it]));
    }
  }
#
# Calls to ODE routine
  cat(sprintf("\n\n ncall = %5d\n\n",ncall));
#
# Plot PDE solutions
#
# u1
  par(mfrow=c(1,1));
  matplot(x=z[2:nz],y=u1[2:nz,],type="l",xlab="z",ylab="u1(z,t)",
          xlim=c(zl,zu),lty=1,main="",lwd=2,col="black");
  persp(z,tout,u1,theta=60,phi=45,
        xlim=c(zl,zu),ylim=c(t0,tf),zlim=c(0,1.1),
        xlab="z",ylab="t",zlab="u1(z,t)");
#
# u2
  par(mfrow=c(1,1));
  matplot(x=z,y=u2,type="l",xlab="z",ylab="u2(z,t)",
          xlim=c(zl,zu),lty=1,main="",lwd=2,col="black");
```

```
  persp(z,tout,u2,theta=60,phi=45,
        xlim=c(zl,zu),ylim=c(t0,tf),zlim=c(0,1.1),
        xlab="z",ylab="t",zlab="u2(z,t)");
#
# u3
  par(mfrow=c(1,1));
  matplot(x=z,y=u3,type="l",xlab="z",ylab="u3(z,t)",
          xlim=c(zl,zu),lty=1,main="",lwd=2,col="black");
  persp(z,tout,u3,theta=60,phi=45,
        xlim=c(zl,zu),ylim=c(t0,tf),zlim=c(0,1.1),
        xlab="z",ylab="t",zlab="u3(z,t)");
```

Listing 5.1: Main program for eqs. (3.1), (3.2), (5.1)

We can note the following details about Listing 5.1 (with some repetition of the discussion of Listings 2.1, 4.1 so that the explanation is self-contained).

- Previous workspaces are deleted.

  ```
  #
  #   Three PDE model
  #
  # Delete previous workspaces
    rm(list=ls(all=TRUE))
  ```

- The R ODE integrator library deSolve is accessed [1].

  ```
  #
  # Access ODE integrator
    library("deSolve");
  #
  # Access functions for numerical solution
    setwd("f:/brain hypoxia/chap5");
    source("pde3a.R");
  ```

  Then the directory with the files for the solution of eqs. (3.1), (3.2), (5.1) is designated. Note that setwd (set working directory) uses / rather than the usual \.

- The model parameters are specified numerically.

  ```
  #
  # Parameters
    nz=21;
    rl=1;
    ru=2;
    D2=0.1;
  ```

Three PDE Model

```
km1=0;
km1=10;
vz=1;
u1e=0.75;
kr3=1;
u1n=1;
u2n=1;
u3n=1;
```

where the additional parameters for eqs. (5.1) are
- kr3=1;: k_{r3} in eq. (5.1-1) for the rate of change of the functional neuron cell density, $u_3(z,t)$, from hypoxia, $-k_{r3}(u_{2n} - u_2)$.
- u3n: normalized neuron cell density in IC (5.1-2).

- The constants $r_1 = \dfrac{2}{r_l},\ r_2 = \dfrac{2r_l}{(r_u^2 - r_l^2)}$ are computed for use in the ordinary differential equation/method of lines (ODE/MOL) routine pde3a.

```
#
# Constants
  r1=2/rl;
  r2=(2*rl)/(ru^2-rl^2);
```

- A spatial grid for eqs. (3.1-1), (3.2-1), (5.1-1) is defined with 21 points so that z = 0,1/20=0.05,...,1. The BBB length is a normalized value, $z = z_u = 1$.

```
#
# Spatial grid in z
  zl=0;zu=1;dz=(zu-zl)/(nz-1);dzs=dz^2;
  z=seq(from=zl,to=zu,by=dz);
```

- An interval in t is defined for 21 output points so that tout=0,1/20=0.05, ..., 1. The time scale is normalized with $t_f = 1$ specified as the final time that is considered appropriate, e.g., week, month, year.

```
#
# Independent variable for ODE integration
  t0=0;tf=1;nout=21;
  tout=seq(from=t0,to=tf,by=(tf-t0)/(nout-1));
```

- ICs (3.1-2), (3.2-2), (5.1-2) are implemented (u1n,u2n,u3n are defined previously as parameters).

```
#
# Initial condition (t=0)
  u0=rep(0,3*nz);
```

```
  for(iz in 1:nz){
    u0[iz]      =u1n;
    u0[iz+nz]   =u2n;
    u0[iz+2*nz]=u3n;
  }
  ncall=0;
```

Also, the counter for the calls to pde3a is initialized.

- The system of 3*nz=63 ODEs is integrated by the library integrator lsodes (available in deSolve, [1]). As expected, the inputs to lsodes are the ODE function, pde3a, the IC vector u0, and the vector of output values of t, tout. The length of u0 (63) informs lsodes how many ODEs are to be integrated. func,y,times are reserved names.

```
#
# ODE integration
  out=lsodes(y=u0,times=tout,func=pde3a,
      sparsetype="sparseint",rtol=1e-6,
      atol=1e-6,maxord=5);
  nrow(out)
  ncol(out)
```

nrow,ncol confirm the dimensions of out.

- $u_1(z,t)$, $u_2(z,t)$, $u_3(z,t)$ are placed in matrices for subsequent plotting.

```
#
# Arrays for plotting numerical solution
  u1=matrix(0,nrow=nz,ncol=nout);
  u2=matrix(0,nrow=nz,ncol=nout);
  u3=matrix(0,nrow=nz,ncol=nout);
  for(it in 1:nout){
    for(iz in 1:nz){
      u1[iz,it]=out[it,iz+1];
      u2[iz,it]=out[it,iz+1+nz];
      u3[iz,it]=out[it,iz+1+2*nz];
    }
    u1[1,it]=u1e;
  }
```

The offset +1 is required because the first element of the solution vectors in out is the value of t and the 2 to 64 elements are the 63 values of u_1, u_2, u_3. These dimensions from the preceding calls to nrow,ncol are confirmed in the subsequent output.

- The numerical values of $u_1(z,t)$, $u_2(z,t)$, $u_3(z,t)$ returned by lsodes are displayed. Every fourth value in t and every second value in z appear from by=4,2.

```
#
# Display numerical solution
  iv=seq(from=1,to=nout,by=4);
  for(it in iv){
    cat(sprintf("\n      t       z      u1(z,t)       u2(z,t)
               u3(z,t)\n"));
    iv=seq(from=1,to=nz,by=2);
    for(iz in iv){
      cat(sprintf("%6.2f %6.2f %12.3e %12.3e %12.3e \n",
          tout[it],z[iz],u1[iz,it],u2[iz,it],u3[iz,it]));
    }
  }
```

- The number of calls to pde3a is displayed at the end of the solution.

```
#
# Calls to ODE routine
  cat(sprintf("\n\n ncall = %5d\n\n",ncall));
```

- $u_1(z,t)$, $u_2(z,t)$, $u_3(z,t)$ are plotted in 2D against z and parametrically in t with the R utility matplot and in 3D with R utility persp. par(mfrow=c(1,1)) specifies a 1×1 matrix of plots, that is, one plot on a page.

```
#
# Plot PDE solutions
#
# u1
  par(mfrow=c(1,1));
  matplot(x=z[2:nz],y=u1[2:nz,],type="l",xlab="z",
          ylab="u1(z,t)",xlim=c(zl,zu),lty=1,main="",lwd=2,
          col="black");
  persp(z,tout,u1,theta=60,phi=45,
        xlim=c(zl,zu),ylim=c(t0,tf),zlim=c(0,1.1),
        xlab="z",ylab="t",zlab="u1(z,t)");
#
# u2
  par(mfrow=c(1,1));
  matplot(x=z,y=u2,type="l",xlab="z",ylab="u2(z,t)",
          xlim=c(zl,zu),lty=1,main="",lwd=2,col="black");
  persp(z,tout,u2,theta=60,phi=45,
        xlim=c(zl,zu),ylim=c(t0,tf),zlim=c(0,1.1),
        xlab="z",ylab="t",zlab="u2(z,t)");
#
# u3
  par(mfrow=c(1,1));
```

```
   matplot(x=z,y=u3,type="l",xlab="z",ylab="u3(z,t)",
          xlim=c(zl,zu),lty=1,main="",lwd=2,col="black");
   persp(z,tout,u3,theta=60,phi=45,
         xlim=c(zl,zu),ylim=c(t0,tf),zlim=c(0,1.1),
         xlab="z",ylab="t",zlab="u3(z,t)");
```

The 2D plot of $u_1(z,t)$ does not include the discontinuous change from IC (3.1-2) to BC (3.1-3), x=z[2:nz],y=u1[2:nz,]. This discontinuity is not included in $u_2(z,t)$, $u_3(z,t)$, so the solution at $z = z_l$ is included, x=z,y=u2, x=z,y=u3.

This completes the discussion of the main program for eqs. (3.1), (3.2), (5.1). The ODE/MOL routine pde3a called by lsodes from the main program for the numerical MOL integration of eqs. (3.1), (3.2), (5.1) is discussed in the following section.

ODE/MOL routine

pde3a called in the main program of Listing 5.1 follows:

```
  pde3a=function(t,u,parm){
#
# Function pde3a computes the t derivatives
# of u1(z,t),u2(z,t),u3(z,t)
#
# One vector to three vectors
  u1=rep(0,nz);
  u2=rep(0,nz);
  u3=rep(0,nz);
  for(iz in 1:nz){
    u1[iz]=u[iz];
    u2[iz]=u[iz+nz];
    u3[iz]=u[iz+2*nz];
  }
#
# BC, z=zl
  u1[1]=u1e;
#
# PDEs
  u1t=rep(0,nz);
  for(iz in 1:nz){
    if(iz==1){u1t[1]=0;}
    if(iz>1){
      u1t[iz]=-vz*(u1[iz]-u1[iz-1])/dz-
        r1*km1*(u1[iz]-u2[iz]);}
  }
  u2t=rep(0,nz);
```

```
  for(iz in 1:nz){
    if(iz==1){
      u2t[1]=
        2*D2*(u2[2]-u2[1])/dzs+
        r2*km1*(u1[1]-u2[1]);}
    if(iz==nz){
      u2t[nz]=
        D2*2*(u2[nz-1]-u2[nz])/dzs+
        r2*km1*(u1[nz]-u2[nz]);}
    if((iz>1)&&(iz<nz)){
      u2t[iz]=
        D2*(u2[iz+1]-2*u2[iz]+u2[iz-1])/dzs+
        r2*km1*(u1[iz]-u2[iz]);}
  }
  u3t=rep(0,nz);
  for(iz in 1:nz){
    u3t[iz]=-kr3*(u2n-u2[iz]);
  }
#
# Three vectors to one vector
  ut=rep(0,3*nz);
  for(iz in 1:nz){
    ut[iz]     =u1t[iz];
    ut[iz+nz]  =u2t[iz];
    ut[iz+2*nz]=u3t[iz];
  }
#
# Increment calls to pde3a
  ncall <<- ncall+1;
#
# Return derivative vector
  return(list(c(ut)));
  }
```

Listing 5.2: ODE/MOL routine for eqs. (3.1), (3.2), (5.1)

We can note the following details about Listing 5.2.

- The function is defined.

    ```
    pde3a=function(t,u,parm){
    #
    # Function pde3a computes the t derivatives
    # of u1(z,t),u2(z,t),u3(z,t)
    ```

    t is the current value of t in eqs. (3.1), (3.2), (5.1). u is the 63-vector of ODE/PDE dependent variables. parm is an argument to pass parameters to pde3a (unused, but required in the argument list). The arguments must be listed in the order stated to properly interface with lsodes called in the main program of Listing 5.1. The derivative vector of the LHS of eqs. (3.1-1), (3.2-1), (5.1-1) is calculated and returned to lsodes, as explained subsequently.

- u is placed in three vectors to facilitate the programming of eqs. (3.1), (3.2), (5.1).

```
#
# One vector to three vectors
  u1=rep(0,nz);
  u2=rep(0,nz);
  u3=rep(0,nz);
  for(iz in 1:nz){
    u1[iz]=u[iz];
    u2[iz]=u[iz+nz];
    u3[iz]=u[iz+2*nz];
  }
```

- BC (3.1-3) is programmed.

```
#
# BC, z=zl
  u1[1]=u1e;
```

- Equation (3.1-1) is programmed.

```
#
# PDEs
  u1t=rep(0,nz);
  for(iz in 1:nz){
    if(iz==1){u1t[1]=0;}
    if(iz>1){
      u1t[iz]=-vz*(u1[iz]-u1[iz-1])/dz-
        r1*km1*(u1[iz]-u2[iz]);}
  }
```

This code has additional explanation in the discussion of Listing 4.2. For $z = z_l$, BC (3.1-3) sets the value of $u_1(z=z_l,t)$ and therefore the derivative is set to zero (if(iz==1)u1t[1]=0;) to ensure the boundary value is maintained.

- Equation (3.2-1) is programmed.

```
  u2t=rep(0,nz);
  for(iz in 1:nz){
    if(iz==1){
      u2t[1]=
```

Three PDE Model

```
          2*D2*(u2[2]-u2[1])/dzs+
          r2*km1*(u1[1]-u2[1]);}
    if(iz==nz){
      u2t[nz]=
          D2*2*(u2[nz-1]-u2[nz])/dzs+
          r2*km1*(u1[nz]-u2[nz]);}
    if((iz>1)&&(iz<nz)){
      u2t[iz]=
          D2*(u2[iz+1]-2*u2[iz]+u2[iz-1])/dzs+
          r2*km1*(u1[iz]-u2[iz]);}
  }
```

This code has additional explanation in the discussion of Listing 4.2.
- Equation (5.1-1) is programmed.

```
  u3t=rep(0,nz);
  for(iz in 1:nz){
    u3t[iz]=-kr3*(u2n-u2[iz]);
  }
```

In general, the correspondence of the programming with eqs. (3.1-1), (3.2-1), (5.1-1) is an important feature of the MOL.
- The 63 ODE derivatives are placed in the vector ut for return to lsodes to take the next in t along the solution.

```
#
# Three vectors to one vector
  ut=rep(0,3*nz);
  for(iz in 1:nz){
    ut[iz]      =u1t[iz];
    ut[iz+nz]   =u2t[iz];
    ut[iz+2*nz]=u3t[iz];
  }
```

- The counter for the calls to pde3a is incremented and returned to the main program of Listing 5.1 by <<-.

```
#
# Increment calls to pde3a
  ncall <<- ncall+1;
```

- The vector ut is returned as a list as required by lsodes. c is the R vector utility. The final } concludes pde3a.

```
    #
    # Return derivative vector
      return(list(c(ut)));
    }
```

This completes the discussion of pde3a. The output from the main program of Listing 5.1 and ODE/MOL routine pde3a of Listing 5.2 is considered next.

Numerical, graphical output

The numerical output is in Table 5.1.

[1] 21

[1] 64

t	z	u1(z,t)	u2(z,t)	u3(z,t)
0.00	0.00	7.500e-01	1.000e+00	1.000e+00
0.00	0.10	1.000e+00	1.000e+00	1.000e+00
0.00	0.20	1.000e+00	1.000e+00	1.000e+00
0.00	0.30	1.000e+00	1.000e+00	1.000e+00
0.00	0.40	1.000e+00	1.000e+00	1.000e+00
0.00	0.50	1.000e+00	1.000e+00	1.000e+00
0.00	0.60	1.000e+00	1.000e+00	1.000e+00
0.00	0.70	1.000e+00	1.000e+00	1.000e+00
0.00	0.80	1.000e+00	1.000e+00	1.000e+00
0.00	0.90	1.000e+00	1.000e+00	1.000e+00
0.00	1.00	1.000e+00	1.000e+00	1.000e+00

t	z	u1(z,t)	u2(z,t)	u3(z,t)
0.20	0.00	7.500e-01	9.087e-01	9.888e-01
0.20	0.10	8.927e-01	9.357e-01	9.937e-01
0.20	0.20	9.556e-01	9.682e-01	9.977e-01
0.20	0.30	9.849e-01	9.880e-01	9.993e-01
0.20	0.40	9.960e-01	9.964e-01	9.998e-01
0.20	0.50	9.991e-01	9.991e-01	1.000e+00
0.20	0.60	9.999e-01	9.998e-01	1.000e+00
0.20	0.70	1.000e+00	1.000e+00	1.000e+00
0.20	0.80	1.000e+00	1.000e+00	1.000e+00
0.20	0.90	1.000e+00	1.000e+00	1.000e+00
0.20	1.00	1.000e+00	1.000e+00	1.000e+00

.
.
.

Output for t = 0.4, 0.6, 0.8 removed

```
    .              .
    .              .
    .              .
  t     z      u1(z,t)       u2(z,t)       u3(z,t)
1.00   0.00   7.500e-01     8.212e-01     8.718e-01
1.00   0.10   8.140e-01     8.367e-01     8.933e-01
1.00   0.20   8.494e-01     8.624e-01     9.222e-01
1.00   0.30   8.792e-01     8.895e-01     9.464e-01
1.00   0.40   9.063e-01     9.149e-01     9.646e-01
1.00   0.50   9.303e-01     9.373e-01     9.776e-01
1.00   0.60   9.504e-01     9.558e-01     9.863e-01
1.00   0.70   9.663e-01     9.701e-01     9.920e-01
1.00   0.80   9.779e-01     9.802e-01     9.954e-01
1.00   0.90   9.855e-01     9.864e-01     9.973e-01
1.00   1.00   9.892e-01     9.886e-01     9.979e-01

ncall =    150
```

Table 5.1: Numerical output from Listings 5.1, 5.2

We can note the following details about this output.

- 21 t output points as the first dimension of the solution matrix out from lsodes as programmed in the main program of Listing 5.1 (with nout=21).
- The solution matrix out returned by lsodes has 64 elements as a second dimension. The first element is the value of t. Elements 2 to 64 are $u_1(z,t)$, $u_2(z,t)$, $u_3(z,t)$ from eqs. (3.1), (3.2), (5.1) (for each of the 21 output points).
- The solution is displayed for t=0,1/20=0.05,...,1 as programmed in Listing 5.1 (every fourth value of t is displayed as explained previously).
- The solution is displayed for z=0,1/20=0.05,...,1 as programmed in Listing 5.1 (every second value of z is displayed as explained previously).
- ICs (3.1-2), (3.2-2), (5.1-2) are confirmed ($t=0$).
- BC (3.1-3) is confirmed ($t=0, z=z_l=0$). There is a discontinuous change from the IC ($t=0$) to the BC ($t>0$) which is subsequently smoothed, e.g., $t=0.2$. To improve the appearance of the graphical output (Figure 5.1-1 considered next), the output does not include the solution at $z=z_l=0$ (matplot(x=z[2:nz],y=u1[2:nz,],... in Listing 5.1). Use of Listings 5.1, 5.2 with the solution at $z=z_l=0$ included is left as an exercise.
- The computational effort as indicated by ncall = 150 is modest so that lsodes computed the solution to eqs. (3.1), (3.2), (5.1) efficiently.

The graphical output for $u_1(z,t)$ is the same as in Figures 4.1, and the output for $u_2(z,t)$ is the same as in Figures 4.2 and is not repeated here.

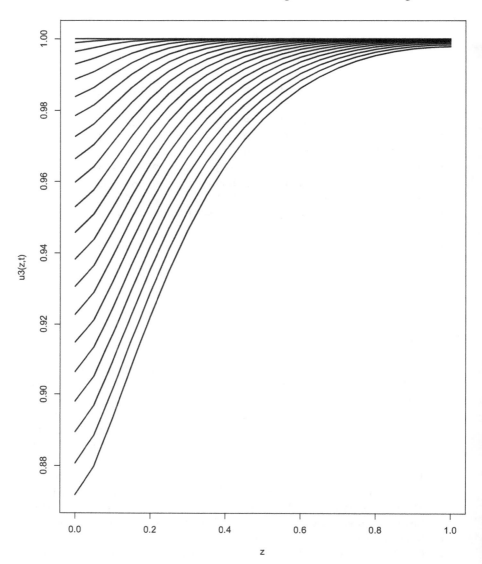

Figure 5.3-1 $u_3(z,t)$ from eqs. (5.1), 2D

The solution $u_3(z,t)$ starts from IC (5.1-2). In general, $u_3(z,t)$ is a front moving left to right (as expected with $v_z > 0$) in response to the reduced entering blood O_2 concentration, $u_1(z = z_l = 0,t) = u_{1e} = 0.75$. The reduced functional neuron cell density from hypoxia might explain the cognitive impairment of long Covid.

Figure 5.3-2 reflects the reduced neuron cell density, $u_3(z,t)$.

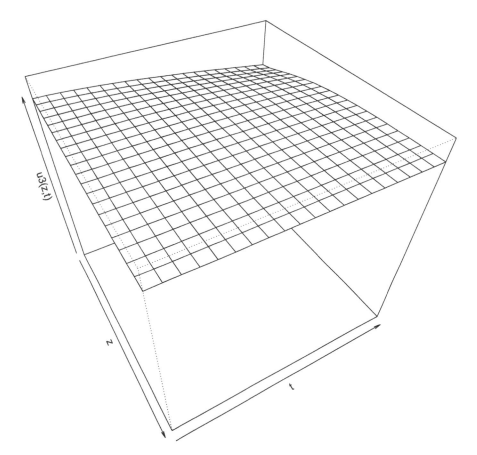

Figure 5.3-2 $u_3(z,t)$ from eqs. (5.1), 3D

5.1.3 SUMMARY AND CONCLUSIONS

Equations (3.1), (3.2), (5.1) are the three PDE model for O_2 in the capillary blood flow, O_2 in the brain tissue, and neuron cell density in the brain tissue. The effect of brain O_2 deficiency (hypoxia) can be studied through the rate constant k_{r3}, which has implications for various neurological effects such as cognitive impairment.

Variation in k_{r3} that might explain recovery from long Covid, and the RHS terms of eqs. (3.1), (3.2), (5.1) are considered in Chapter 6.

REFERENCE

1. Soetaert, K., J. Cash, and F. Mazzia (2012), *Solving Differential Equations in R*, Springer-Verlag, Heidelberg, Germany.

6 Case Studies

6.1 INTRODUCTION

The three PDE model of eqs. (3.1), (3.2), (5.1) is applied to two case studies in this concluding chapter.

6.1.1 TIME VARIATION OF THE BRAIN O_2 CONCENTRATION

The effect of a time variation of the entering O_2 capillary blood concentration, $u_1(z = z_l, t) = u_{1e}(t)$ in boundary condition (3.1-3) can be programmed in the ODE/MOL routine pde3a of Listing 5.2. This variation might, for example, be the result of a partial recovery of the respiratory/lung function with time, e.g., through the use of supplemental O_2.

Main program

The time variation of the entering blood capillary O_2 concentration, $u_{1e}(t)$, is programmed as minor changes/additions to the main program of Listing 5.1 and the ODE/MOL routine of Listing 5.2. The changes/additions to the main program are considered next.

```
     .
     .
     .
#
# Access functions for numerical solution
  setwd("f:/hypoxia/chap6");
  source("pde3a.R");
#
# Set case
  ncase=1;
#
# Parameters
  nz=21;
  rl=1;
  ru=2;
  D2=0.1;
  km1=0;
  km1=10;
  vz=1;
```

```
    u1e=0.75;
    kr3=1;
    u1n=1;
    u2n=1;
    u3n=1;
         .
         .
         .
    if(ncase==1){u1[1,it]=u1e;}
    if(ncase==2){u1[1,it]=u1n+(u1e-u1n)*sin(pi*tout[it]);}
    if(ncase==3){
      if(tout[it]<=0.5){u1[1,it]=u1n+(u1e-u1n)*sin(pi*tout[it]);}
      if(tout[it]> 0.5){u1[1,it]=u1e;}
         .
         .
         .
#
# Plot PDE solutions
#
# u1
  par(mfrow=c(1,1));
  matplot(x=z[2:nz],y=u1[2:nz,],type="l",xlab="z",ylab="u1(z,t)",
          xlim=c(zl,zu),lty=1,main="",lwd=2,col="black");
  persp(z,tout,u1,theta=120,phi=25,
        xlim=c(zl,zu),ylim=c(t0,tf),
        xlab="z",ylab="t",zlab="u1(z,t)");
#
# u2
  par(mfrow=c(1,1));
  matplot(x=z,y=u2,type="l",xlab="z",ylab="u2(z,t)",
          xlim=c(zl,zu),lty=1,main="",lwd=2,col="black");
  persp(z,tout,u2,theta=120,phi=25,
        xlim=c(zl,zu),ylim=c(t0,tf),
        xlab="z",ylab="t",zlab="u2(z,t)");
#
# u3
  par(mfrow=c(1,1));
  matplot(x=z,y=u3,type="l",xlab="z",ylab="u3(z,t)",
          xlim=c(zl,zu),lty=1,main="",lwd=2,col="black");
  persp(z,tout,u3,theta=120,phi=25,
        xlim=c(zl,zu),ylim=c(t0,tf),
        xlab="z",ylab="t",zlab="u3(z,t)");
#
# u1e(t)=u1(z=z_l,t)
```

Case Studies

```
par(mfrow=c(1,1));
matplot(x=tout,y=u1[1,],type="l",xlab="t",ylab="u1e(t)",
        xlim=c(t0,tf),lty=1,main="",lwd=2,col="black");
```

Listing 6.1: Changes/additions to the main program for $u_1(z = z_l, t) = u_{1e}(t)$

We can note the following details about Listing 6.1 (refer also to Listing 5.1).

- The ODE/MOL routine is again pde3a (considered next).

  ```
  #
  # Access functions for numerical solution
    setwd("f:/hypoxia/chap6");
    source("pde3a.R");
  ```

- A case index, ncase, is included, with ncase=1,2,3 as discussed subsequently.

  ```
  #
  # Set case
    ncase=1;
  #
  # Parameters
    nz=21;
    rl=1;
    ru=2;
    D2=0.1;
    km1=0;
    km1=10;
    vz=1;
    u1e=0.75;
    kr3=1;
    u1n=1;
    u2n=1;
    u3n=1;
  ```

  The parameters have the same numerical values as given in Listing 5.1.

- BC (3.1-3) is programmed for three values of ncase.

  ```
  if(ncase==1){u1[1,it]=u1e;}
  if(ncase==2){u1[1,it]=u1n+(u1e-u1n)*sin(pi*tout[it]);}
  if(ncase==3){
    if(tout[it]<=0.5){u1[1,it]=u1n+(u1e-u1n)
                                    *sin(pi*tout[it]);}
    if(tout[it]> 0.5){u1[1,it]=u1e;}
  ```

  For ncase=1, the entering blood O_2 concentration is the constant u_{1e} = u1e (the same as in Listing 5.1).

For ncase=2, the entering blood O_2 concentration change is a half sine wave, as explained subsequently (after the pde3a changes, Listing 6.2). This could represent an initial impairment (injury, damage) of the respiratory/lung function, followed by recovery.

For ncase=3, the entering blood O_2 concentration change is a half sine wave, followed by a permanent decrease, as explained subsequently (after the pde3a changes, Listing 6.2). This could represent an initial impairment (injury, damage) of the respiratory/lung function, without recovery.

- The arguments of persp were modified to enhance the appearance of the 3D plots. For example, for the 3D plotting of $u_1(z,t)$,

Listing 5.1

```
persp(z,tout,u1,theta=60,phi=45,
      xlim=c(zl,zu),ylim=c(t0,tf),zlim=c(0,1.1),
      xlab="z",ylab="t",zlab="u1(z,t)");
```

Listing 6.1

```
persp(z,tout,u1,theta=120,phi=25,
      xlim=c(zl,zu),ylim=c(t0,tf),
      xlab="z",ylab="t",zlab="u1(z,t)");
```

These changes have the effect of reversing the direction of the z axis in the 3D plots.

- The time variation in $u_{1e}(t)$ is plotted against t at the end of the main program to confirm the variation for ncase=1,2,3.

```
#
# u1e(t)=u1(z=z_l,t)
  par(mfrow=c(1,1));
  matplot(x=tout,y=u1[1,],type="l",xlab="t",ylab="u1e(t)",
          xlim=c(t0,tf),lty=1,main="",lwd=2,col="black");
```

This concludes the discussion of the changes/additions to the main program of Listing 5.1 for $u_{1e}(t)$ = u1e. The ODE/MOL routine, pde3a, is considered next.

ODE/MOL routine

The additions to pde3a are for BC (3.1-3) with ncase=1,2,3.

```
      .
      .
      .
#
# BC, z=zl
```

```
if(ncase==1){u1[1]=u1e;}
if(ncase==2){u1[1]=u1n+(u1e-u1n)*sin(pi*t);}
if(ncase==3){
  if(t<=0.5){u1[1]=u1n+(u1e-u1n)*sin(pi*t);}
  if(t>0.5 ){u1[1]=u1e;}
}
```

.
.
.

Listing 6.2: Additions to the ODE/MOL routine of Listing 5.2

We can note the following details about Listing 6.2 (refer also to Listing 5.2).

- For ncase=1, the entering blood O_2 concentration is the constant u_{1e} = u1e (the same as in Listing 5.2).

  ```
  #
  # BC, z=zl
    if(ncase==1){u1[1]=u1e;}
  ```

- For ncase=2, the entering blood O_2 concentration $u_{1e}(t)$ is a half sine wave that passes through the points
  (1) $u_{1e}(t = t_0 = 0) = u_{1n}$ (u_{1n} is the IC (3.1-2)).
  (2) $u_{1e}(t = (t_f - t_0)/2 = 0.5) = u_{1e}$
  (3) $u_{1e}(t = t_f = 1) = u_{1n}$

  ```
  if(ncase==2){u1[1]=u1n+(u1e-u1n)*sin(pi*t);}
  ```

  This variation in $u_{1e}(t) = u_1(z = z_l, t)$ could result, for example, from initial impairment of the respiratory/lung system, followed by recovery.

- For ncase=3, the entering blood O_2 concentration $u_{1e}(t)$ is a half sine wave that passes through the points
  (1) $u_{1e}(t = t_0 = 0) = u_{1n}$ (u_{1n} is the IC (3.1-2)).
  (2) $u_{1e}(t = (t_f - t_0)/2 = 0.5) = u_{1e}$
  (3) $u_{1e}(t = t_f = 1) = u_{1e}$

  ```
  if(ncase==3){
    if(t<=0.5){u1[1]=u1n+(u1e-u1n)*sin(pi*t);}
    if(t>0.5 ){u1[1]=u1e;}
  }
  ```

  This variation in $u_{1e}(t) = u_1(z = z_l, t)$ could result, for example, from initial impairment of the respiratory/lung system, without recovery.

The numerical and graphical output from the main program of Listings 5.1, 6.1, and the ODE/MOL routine, pde3a, of Listings 5.2, 6.2 follows.

Numerical and graphical output

For ncase=1 in Listing 6.1, the numerical output is the same as given in Table 5.1, so this solution is not repeated here. The graphical $u_{1e}(t)$ (set as a parameter and plotted in Listing 6.1) is the constant 0.75 (invariant in t) and is not presented here.

For ncase=2 in Listing 6.1, the numerical output is given as follows:

[1] 21

[1] 64

t	z	u1(z,t)	u2(z,t)	u3(z,t)
0.00	0.00	1.000e+00	1.000e+00	1.000e+00
0.00	0.10	1.000e+00	1.000e+00	1.000e+00
0.00	0.20	1.000e+00	1.000e+00	1.000e+00
0.00	0.30	1.000e+00	1.000e+00	1.000e+00
0.00	0.40	1.000e+00	1.000e+00	1.000e+00
0.00	0.50	1.000e+00	1.000e+00	1.000e+00
0.00	0.60	1.000e+00	1.000e+00	1.000e+00
0.00	0.70	1.000e+00	1.000e+00	1.000e+00
0.00	0.80	1.000e+00	1.000e+00	1.000e+00
0.00	0.90	1.000e+00	1.000e+00	1.000e+00
0.00	1.00	1.000e+00	1.000e+00	1.000e+00

t	z	u1(z,t)	u2(z,t)	u3(z,t)
0.20	0.00	8.531e-01	9.662e-01	9.974e-01
0.20	0.10	9.600e-01	9.808e-01	9.988e-01
0.20	0.20	9.893e-01	9.929e-01	9.996e-01
0.20	0.30	9.975e-01	9.979e-01	9.999e-01
0.20	0.40	9.995e-01	9.995e-01	1.000e+00
0.20	0.50	9.999e-01	9.999e-01	1.000e+00
0.20	0.60	1.000e+00	1.000e+00	1.000e+00
0.20	0.70	1.000e+00	1.000e+00	1.000e+00
0.20	0.80	1.000e+00	1.000e+00	1.000e+00
0.20	0.90	1.000e+00	1.000e+00	1.000e+00
0.20	1.00	1.000e+00	1.000e+00	1.000e+00

.
.
.

Output for t = 0.4, 0.6, 0.8 removed

.
.
.

t	z	u1(z,t)	u2(z,t)	u3(z,t)
1.00	0.00	1.000e+00	9.108e-01	9.144e-01
1.00	0.10	9.139e-01	9.036e-01	9.284e-01

```
1.00    0.20    9.018e-01    9.065e-01    9.484e-01
1.00    0.30    9.126e-01    9.201e-01    9.656e-01
1.00    0.40    9.309e-01    9.381e-01    9.783e-01
1.00    0.50    9.500e-01    9.557e-01    9.870e-01
1.00    0.60    9.663e-01    9.704e-01    9.926e-01
1.00    0.70    9.787e-01    9.813e-01    9.960e-01
1.00    0.80    9.872e-01    9.887e-01    9.979e-01
1.00    0.90    9.923e-01    9.929e-01    9.988e-01
1.00    1.00    9.948e-01    9.943e-01    9.991e-01

ncall =    124
```

Table 6.2: Numerical output from Listings 5.1, 5.2, 6.1, 6.2, ncase=2

We can note the following details about this output.

- 21 t output points as the first dimension of the solution matrix out from lsodes as programmed in the main program of Listing 5.1 (with nout=21).
- The solution matrix out returned by lsodes has 64 elements as a second dimension. The first element is the value of t. Elements 2 to 64 are $u_1(z,t)$, $u_2(z,t)$, $u_3(z,t)$ from eqs. (3.1), (3.2), (5.1) (for each of the 21 output points).
- The solution is displayed for t=0,1/20=0.05,...,1 as programmed in Listing 5.1 (every fourth value of t is displayed as explained previously).
- The solution is displayed for z=0,1/20=0.05,...,1 as programmed in Listing 5.1 (every second value of z is displayed as explained previously).
- ICs (3.1-2), (3.2-2), (5.1-2) are confirmed ($t = 0$).
- BC (3.1-3) is confirmed ($u_1(z = z_l = 0, t) = u_{1e}(t)$) according to the half sine wave as programmed in Listings 6.1, 6.2 for ncase=2.
- The computational effort as indicated by ncall = 124 is modest so that lsodes computed the solution to eqs. (3.1), (3.2), (5.1) efficiently.

The graphical output is shown in Figures 6.1, 2, 3, 4.

Figure 6.1-1 indicates the transient of $u_1(z,t)$ away from IC (3.1-2), $u_1(z, t = 0) = u_{1n} = 1$ in response to BC (3.1-3), $u_1(z = z_l, t) = u_{1e}(t)$.

Figure 6.1-2 confirms the response in z, t of Figure 6.1-1. The discontinuity at $z = z_l = 0$ is not included in Figure 6.1-1, but is included in Figure 6.1-2, as discussed previously.

Figure 6.2-1 indicates the transient of $u_2(z,t)$ away from IC (3.2-2), in response to the O_2 transfer through the BBB. Figure 6.2-2 confirms the response in z, t of Figure 6.2-1.

Figure 6.3-1 indicates the continuing decrease in the neuron cell density according to eq. (5.1-1).

Figure 6.3-2 confirms the response in z, t of Figure 6.3-1.

Figure 6.4-1 confirms $u_{1e}(t)$ programmed in Listings 6.1, 6.2 for ncase=2. In general, the numerical and graphical output indicates modest changes in

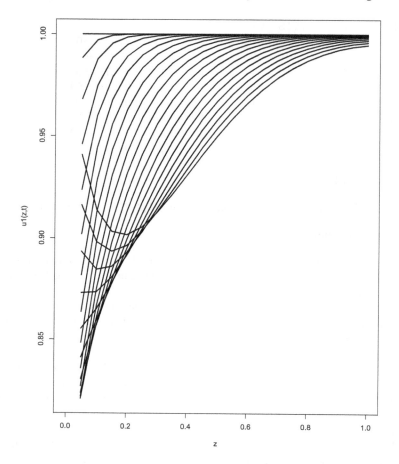

Figure 6.1-1 $u_1(z,t)$ from eqs. (3.1), 2D, ncase=2

$u_1(z,t)$, $u_2(z,t)$, $u_3(z,t)$ resulting from the recovery of $u_{1e}(t)$ with increasing t (Figure 6.4-1).

For ncase3, $u_{1e}(t)$ does not recover with increasing t so that the changes in $u_1(z,t)$, $u_2(z,t)$, $u_3(z,t)$ are larger than for ncase=2. This is confirmed by the following output (for ncase=3 in Listings 6.1, 6.2).

```
[1] 21

[1] 64

     t      z    u1(z,t)      u2(z,t)      u3(z,t)
  0.00   0.00  1.000e+00    1.000e+00    1.000e+00
  0.00   0.10  1.000e+00    1.000e+00    1.000e+00
  0.00   0.20  1.000e+00    1.000e+00    1.000e+00
```

```
  0.00   0.30   1.000e+00   1.000e+00   1.000e+00
  0.00   0.40   1.000e+00   1.000e+00   1.000e+00
  0.00   0.50   1.000e+00   1.000e+00   1.000e+00
  0.00   0.60   1.000e+00   1.000e+00   1.000e+00
  0.00   0.70   1.000e+00   1.000e+00   1.000e+00
  0.00   0.80   1.000e+00   1.000e+00   1.000e+00
  0.00   0.90   1.000e+00   1.000e+00   1.000e+00
  0.00   1.00   1.000e+00   1.000e+00   1.000e+00

    t      z      u1(z,t)     u2(z,t)     u3(z,t)
  0.20   0.00   8.531e-01   9.662e-01   9.974e-01
  0.20   0.10   9.600e-01   9.808e-01   9.988e-01
  0.20   0.20   9.893e-01   9.929e-01   9.996e-01
  0.20   0.30   9.975e-01   9.979e-01   9.999e-01
  0.20   0.40   9.995e-01   9.995e-01   1.000e+00
  0.20   0.50   9.999e-01   9.999e-01   1.000e+00
  0.20   0.60   1.000e+00   1.000e+00   1.000e+00
  0.20   0.70   1.000e+00   1.000e+00   1.000e+00
  0.20   0.80   1.000e+00   1.000e+00   1.000e+00
  0.20   0.90   1.000e+00   1.000e+00   1.000e+00
  0.20   1.00   1.000e+00   1.000e+00   1.000e+00
                      .           .
                      .           .
                      .           .
  Output for t = 0.4, 0.6, 0.8 removed
                      .           .
                      .           .
                      .           .

    t      z      u1(z,t)     u2(z,t)     u3(z,t)
  1.00   0.00   7.500e-01   8.325e-01   9.029e-01
  1.00   0.10   8.242e-01   8.500e-01   9.213e-01
  1.00   0.20   8.644e-01   8.784e-01   9.452e-01
  1.00   0.30   8.967e-01   9.071e-01   9.643e-01
  1.00   0.40   9.245e-01   9.326e-01   9.778e-01
  1.00   0.50   9.475e-01   9.536e-01   9.868e-01
  1.00   0.60   9.654e-01   9.697e-01   9.925e-01
  1.00   0.70   9.784e-01   9.811e-01   9.959e-01
  1.00   0.80   9.871e-01   9.886e-01   9.979e-01
  1.00   0.90   9.923e-01   9.929e-01   9.988e-01
  1.00   1.00   9.948e-01   9.943e-01   9.991e-01

ncall =    153
```

Table 6.3: Numerical output from Listings 5.1, 5.2, 6.1, 6.2, ncase=3

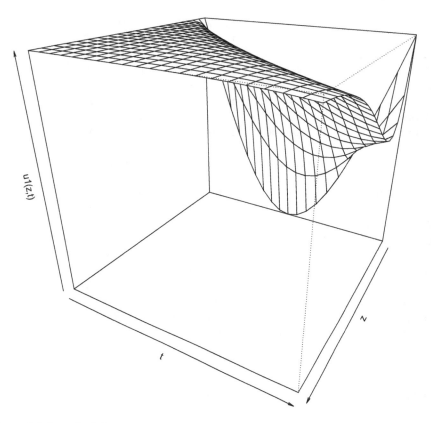

Figure 6.1-2 $u_1(z,t)$ from eqs. (3.1), 3D, `ncase=2`

We can note the following details about this output.

- The dimensions of array out from `lsodes` are the same as for `ncase=2` (Table 6.2).
- ICs (3.1-2), (3.2-2), (5.1-2) are confirmed ($t=0$).
- BC (3.1-3) is confirmed ($u_1(z=z_l=0,t)=u_{1e}(t)$) according to the half sine wave as programmed in Listings 6.1, 6.2 for `ncase=3` (e.g., $u_1(z=0, t=1)=0.75$).
- The computational effort as indicated by `ncall = 153` is modest so that `lsodes` computed the solution to eqs. (3.1), (3.2), (5.1) efficiently.

The graphical output is shown in Figures 6.5,6,7,8.

Figure 6.5-1 indicates the monotonic decrease of the capillary blood O_2 concentration toward 0.75 with increasing t.

Figure 6.5-2 confirms the response in z, t of Figure 6.5-1.

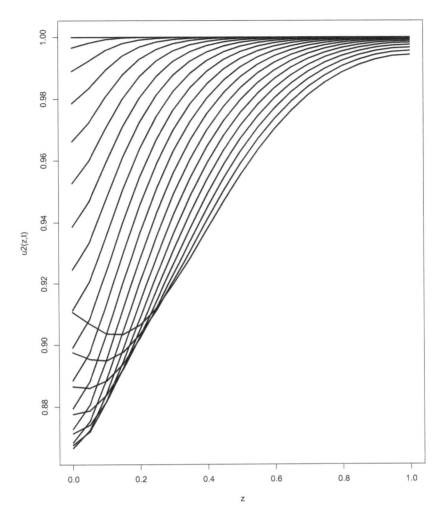

Figure 6.2-1 $u_2(z,t)$ from eqs. (3.2), 2D, ncase=2

Figure 6.6-1 indicates the monotonic decrease of the brain tissue O_2 concentration with increasing t in response to the O_2 through the BBB. Figure 6.6-2 confirms the response in z, t of Figure 6.6-1.

Figure 6.7-1 indicates the continuing decrease in the neuron cell density according to eq. (5.1-1).

Figure 6.7-2 confirms the response in z, t of Figure 6.7-1.

Figure 6.8-1 confirms $u_{1e}(t)$ programmed in Listings 6.1, 6.2 for ncase=3.

In general, the numerical and graphical output indicates lower O_2 concentrations and neuron cell density for ncase=3 than for ncase=2. For example,

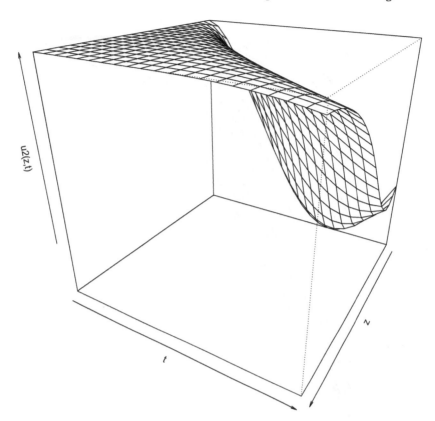

Figure 6.2-2 $u_2(z,t)$ from eqs. (3.1), 3D, ncase=2

Table 6.2, ncase=2

t	z	u1(z,t)	u2(z,t)	u3(z,t)
1.00	0.10	9.139e-01	9.036e-01	9.284e-01

Table 6.3, ncase=3

t	z	u1(z,t)	u2(z,t)	u3(z,t)
1.00	0.10	8.242e-01	8.500e-01	9.213e-01

The reduced O_2 concentrations and neuron cell density could result in cognitive impairment originating from Covid-reduced respiratory/lung function.

A similar analysis can be carried out with a variable blood flow rate, $v_z(t)$, which could also result from a Covid-reduced respiratory/lung function. This is left as an exercise.

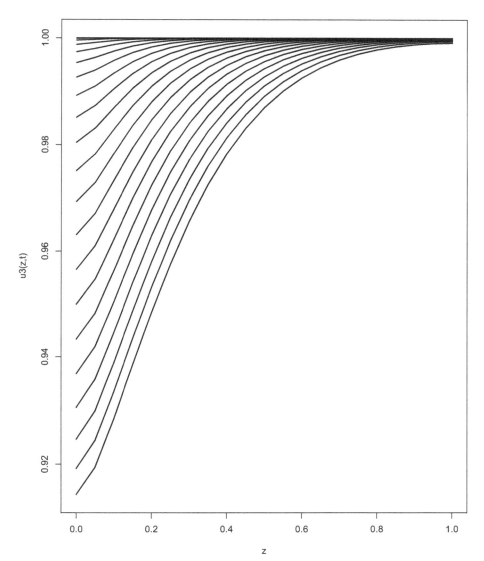

Figure 6.3-1 $u_3(z,t)$ from eqs. (5.1), 2D, ncase=2

Additionally, the effect of the discontinuity between IC (3.1-2), $u_{(z\,=\,z_l\,=\,0,t\,=\,0)} = u_{1n} = 1$, and BC (3.1-3), $u_{(z\,=\,z_l\,=\,0,t\,>\,0)} = u_{1e} = 0.75$, can be observed by changing the plotting of u_1 from

```
matplot(x=z[2:nz],y=u1[2:nz,],type="l",xlab="z",ylab="u1(z,t)",
        xlim=c(zl,zu),lty=1,main="",lwd=2,col="black");
```

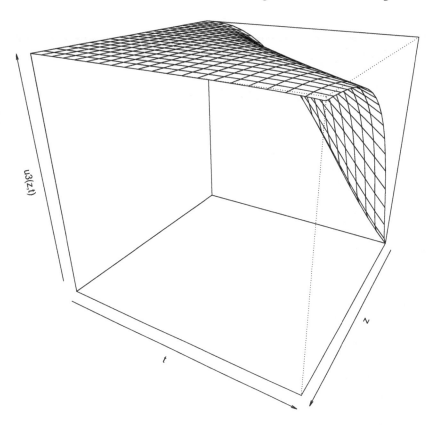

Figure 6.3-2 $u_3(z,t)$ from eqs. (5.1), 3D, `ncase=2`

to

```
matplot(x=z,y=u1,type="l",xlab="z",ylab="u1(z,t)",
    xlim=c(zl,zu),lty=1,main="",lwd=2,col="black");
```

This is left as an exercise.

The complex solutions for $u_1(z,t)$, $u_2(z,t)$, $u_3(z,t)$, as reflected in Figures 6.1,...,8, are a result of the integration of the left hand side (LHS) t derivatives of eqs. (3.1-1), (3.2-1), (5.1-1). These derivatives can be computed from the solutions and displayed as explained in the next case.

6.1.2 LHS PDE TIME DERIVATIVES

The LHS t derivatives are computed, then displayed, with R code added to the main programs of Listings 5.1, 6.1. The ODE/MOL routine is again `pde3a` in Listings 5.2, 6.2.

Case Studies

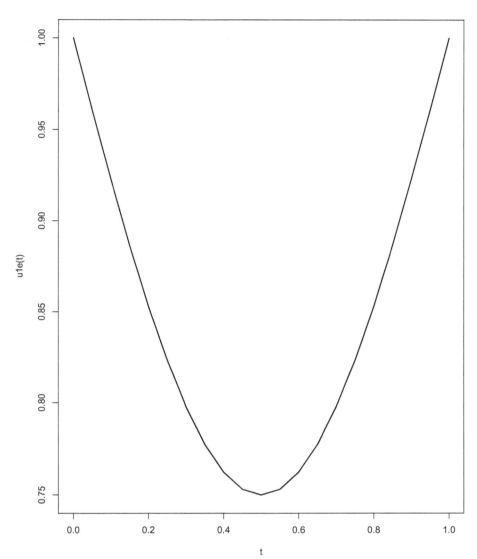

Figure 6.4-1 $u_{1e}(t)$ from eq. (3.1-1), `ncase=2`

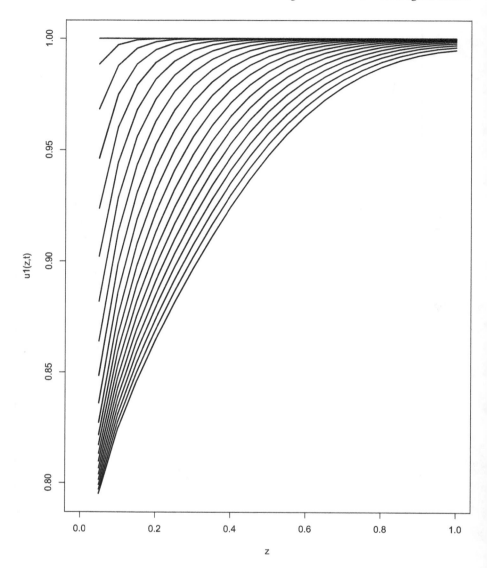

Figure 6.5-1 $u_1(z,t)$ from eqs. (3.1), 2D, `ncase=3`

Case Studies

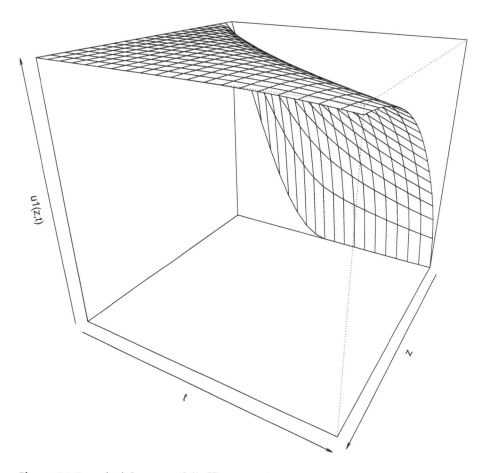

Figure 6.5-2 $u_1(z,t)$ from eqs. (3.1), 3D, `ncase=3`

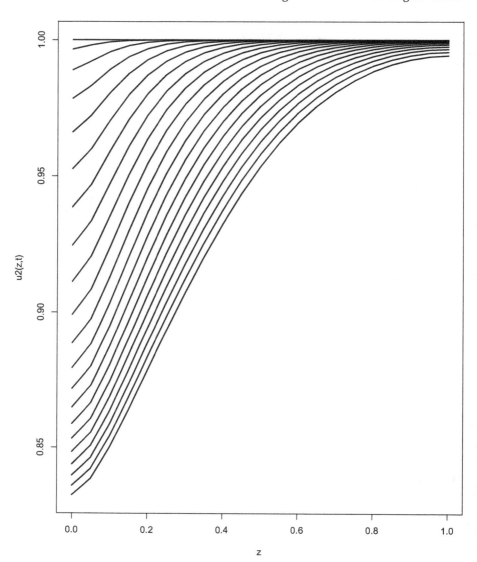

Figure 6.6-1 $u_2(z,t)$ from eqs. (3.2), 2D, `ncase=3`

Case Studies

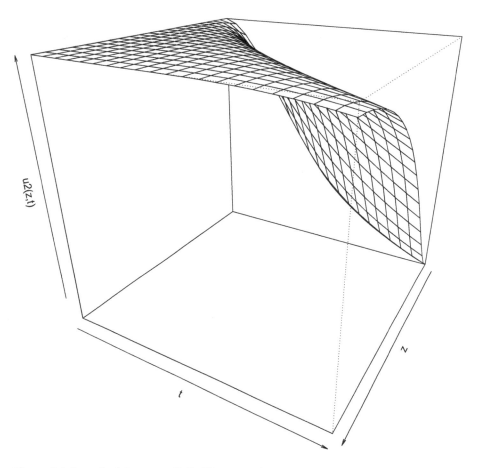

Figure 6.6-2 $u_2(z,t)$ from eqs. (3.2), 3D, `ncase=3`

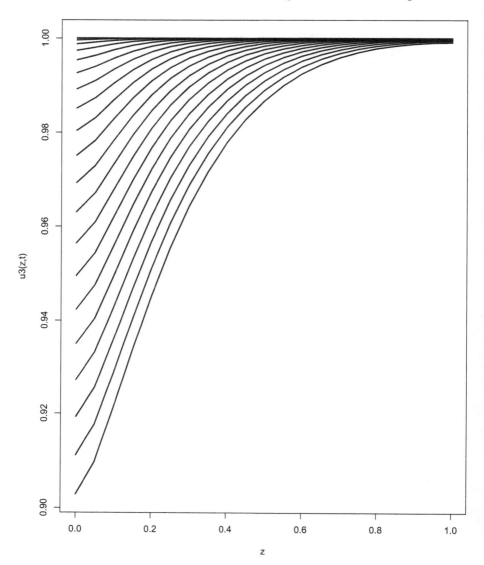

Figure 6.7-1 $u_3(z,t)$ from eqs. (5.1), 2D, `ncase=3`

Case Studies

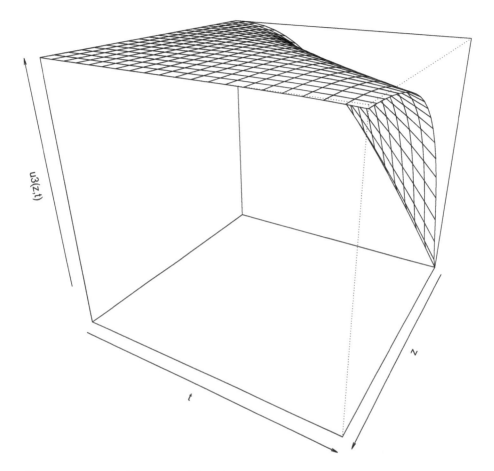

Figure 6.7-2 $u_3(z,t)$ from eqs. (5.1), 3D, `ncase=3`

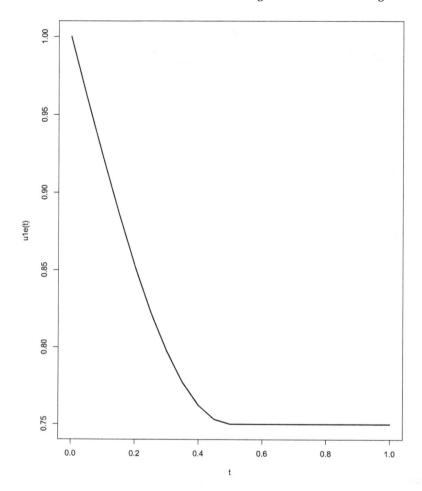

Figure 6.8-1 $u_{1e}(t)$ from eq. (3.1-1), ncase=3

Case Studies

Main program

The following code is added to the end of the main programs of Listings 5.1, 6.1 to compute and plot $\dfrac{\partial u_1(z,t)}{\partial t}$, $\dfrac{\partial u_2(z,t)}{\partial t}$, $\dfrac{\partial u_3(z,t)}{\partial t}$ from eqs. (3.1-1), (3.2-1), (5.1-1).

```
  .
  .
  .
  Main program from Listings 5.1, 6.1
  .
  .
  .
#
# Derivatives in t
  u1t=matrix(0,nrow=nz,ncol=nout);
  u2t=matrix(0,nrow=nz,ncol=nout);
  u3t=matrix(0,nrow=nz,ncol=nout);
#
# Derivative calculation
#
  for(it in 1:nout){
#
# u1t
  for(iz in 1:nz){
    if(iz==1){u1t[1,it]=0;}
    if(iz>1){
      u1t[iz,it]=-vz*(u1[iz,it]-u1[iz-1,it])/dz-
        r1*km1*(u1[iz,it]-u2[iz,it]);}
  }
#
# u2t
  for(iz in 1:nz){
    if(iz==1){
      u2t[1,it]=
        2*D2*(u2[2,it]-u2[1,it])/dzs+
        r2*km1*(u1[1,it]-u2[1,it]);}
    if(iz==nz){
      u2t[nz,it]=
        D2*2*(u2[nz-1,it]-u2[nz,it])/dzs+
        r2*km1*(u1[nz,it]-u2[nz,it]);}
    if((iz>1)&&(iz<nz)){
      u2t[iz,it]=
        D2*(u2[iz+1,it]-2*u2[iz,it]+u2[iz-1,it])/dzs+
        r2*km1*(u1[iz,it]-u2[iz,it]);}
  }
```

```
#
# u3t
  for(iz in 1:nz){
    u3t[iz,it]=-kr3*(u2n-u2[iz,it]);
  }
  }
#
# Derivative plotting
#
# u1t
  par(mfrow=c(1,1));
  matplot(x=z[2:nz],y=u1t[2:nz,],type="l",xlab="z",
          ylab="u1t(z,t)",xlim=c(zl,zu),lty=1,main="",lwd=2,
          col="black");
  persp(z,tout,u1t,theta=120,phi=25,
        xlim=c(zl,zu),ylim=c(t0,tf),
        xlab="z",ylab="t",zlab="u1t(z,t)");
#
# u2t
  par(mfrow=c(1,1));
  matplot(x=z,y=u2t,type="l",xlab="z",ylab="u2t(z,t)",
          xlim=c(zl,zu),lty=1,main="",lwd=2,col="black");
  persp(z,tout,u2t,theta=120,phi=25,
        xlim=c(zl,zu),ylim=c(t0,tf),
        xlab="z",ylab="t",zlab="u2t(z,t)");
#
# u3t
  par(mfrow=c(1,1));
  matplot(x=z,y=u3t,type="l",xlab="z",ylab="u3t(z,t)",
          xlim=c(zl,zu),lty=1,main="",lwd=2,col="black");
  persp(z,tout,u3t,theta=120,phi=25,
        xlim=c(zl,zu),ylim=c(t0,tf),
        xlab="z",ylab="t",zlab="u3t(z,t)");
```

Listing 6.3: Additions to the main programs of Listings 5.1, 6.1 to compute and plot the t derivatives

We can note the following details about Listing 6.3.

- Matrices are defined for $\frac{\partial u_1(z,t)}{\partial t}, \frac{\partial u_2(z,t)}{\partial t}, \frac{\partial u_3(z,t)}{\partial t}$.

```
#
# Derivatives in t
  u1t=matrix(0,nrow=nz,ncol=nout);
  u2t=matrix(0,nrow=nz,ncol=nout);
  u3t=matrix(0,nrow=nz,ncol=nout);
```

nz,nout are defined numerically in Listing 5.1.

Case Studies 79

- The coding for $\dfrac{\partial u_1(z,t)}{\partial t}$ of eq. (3.1-1) is taken from pde3a of Listing 5.2 with a subscript it for t added.

```
#
# Derivative calculation
#
  for(it in 1:nout){
#
# u1t
  for(iz in 1:nz){
    if(iz==1){u1t[1,it]=0;}
    if(iz>1){
      u1t[iz,it]=-vz*(u1[iz,it]-u1[iz-1,it])/dz-
        r1*km1*(u1[iz,it]-u2[iz,it]);}
  }
```

- The coding for $\dfrac{\partial u_2(z,t)}{\partial t}$ of eq. (3.2-1) is taken from pde3a of Listing 5.2 with a subscript it for t added.

```
#
# u2t
  for(iz in 1:nz){
    if(iz==1){
      u2t[1,it]=
        2*D2*(u2[2,it]-u2[1,it])/dzs+
        r2*km1*(u1[1,it]-u2[1,it]);}
    if(iz==nz){
      u2t[nz,it]=
        D2*2*(u2[nz-1,it]-u2[nz,it])/dzs+
        r2*km1*(u1[nz,it]-u2[nz,it]);}
    if((iz>1)&&(iz<nz)){
      u2t[iz,it]=
        D2*(u2[iz+1,it]-2*u2[iz,it]+u2[iz-1,it])/dzs+
        r2*km1*(u1[iz,it]-u2[iz,it]);}
  }
```

- The coding for $\dfrac{\partial u_3(z,t)}{\partial t}$ of eq. (5.1-1) is taken from pde3a of Listing 5.2 with a subscript it for t added.

```
#
# u3t
  for(iz in 1:nz){
    u3t[iz,it]=-kr3*(u2n-u2[iz,it]);
  }
  }
```

The second } concludes the loop in it.

- The plotting of $\dfrac{\partial u_1(z,t)}{\partial t}$, $\dfrac{\partial u_2(z,t)}{\partial t}$, $\dfrac{\partial u_3(z,t)}{\partial t}$ as a function of z,t is with matplot (2D) and persp (3D).

```
#
# Derivative plotting
#
# u1t
  par(mfrow=c(1,1));
  matplot(x=z[2:nz],y=u1t[2:nz,],type="l",xlab="z",
          ylab="u1t(z,t)",xlim=c(zl,zu),lty=1,main="",lwd=2,
          col="black");
  persp(z,tout,u1t,theta=120,phi=25,
        xlim=c(zl,zu),ylim=c(t0,tf),
        xlab="z",ylab="t",zlab="u1t(z,t)");
#
# u2t
  par(mfrow=c(1,1));
  matplot(x=z,y=u2t,type="l",xlab="z",ylab="u2t(z,t)",
          xlim=c(zl,zu),lty=1,main="",lwd=2,col="black");
  persp(z,tout,u2t,theta=120,phi=25,
        xlim=c(zl,zu),ylim=c(t0,tf),
        xlab="z",ylab="t",zlab="u2t(z,t)");
#
# u3t
  par(mfrow=c(1,1));
  matplot(x=z,y=u3t,type="l",xlab="z",ylab="u3t(z,t)",
          xlim=c(zl,zu),lty=1,main="",lwd=2,col="black");
  persp(z,tout,u3t,theta=120,phi=25,
        xlim=c(zl,zu),ylim=c(t0,tf),
        xlab="z",ylab="t",zlab="u3t(z,t)");
```

In summary, the main program consists of Listings 5.1, 6.1, 6.3.

ODE/MOL routine

The ODE/MOL routine is the same as given in Listings 5.2, 6.2.

Numerical and graphical output

The graphical output from Listings 5.1, 6.1, 6.3 for $\dfrac{\partial u_1(z,t)}{\partial t}$, $\dfrac{\partial u_2(z,t)}{\partial t}$, $\dfrac{\partial u_3(z,t)}{\partial t}$ follows (the numerical and graphical output for $u_1(z,t), u_2(z,t), u_2(z,t)$ was presented previously in Tables 6.2, 6.3 and Figures 6.1-1,...,8).

Figure 6.9-1 indicates $\dfrac{\partial u_1}{\partial t} < 0$ with a reversal in $u_{1e}(t)$ at $t = 0.5$.

Figure 6.9-2 confirms Figure 6.9-1.

Figure 6.10-1 indicates $\dfrac{\partial u_2}{\partial t} < 0$ with a reversal in $u_{1e}(t)$ at $t = 0.5$.

Case Studies

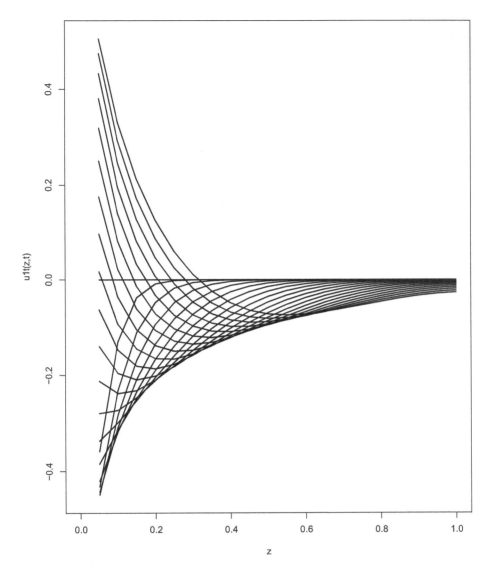

Figure 6.9-1 $\dfrac{\partial u_1(z,t)}{\partial t}$ from eqs. (3.1), 2D, `ncase=2`

Figure 6.10-2 confirms Figure 6.10-1.

Figure 6.11-1 indicates $\dfrac{\partial u_3}{\partial t} < 0$ and continuing decrease of u_3 with t. This decrease in $u_3(z,t)$ results from the single RHS term in eq. (5.1-1), $-k_{r3}(u_{2n} - u_2) < 0$.

Figure 6.11-2 confirms Figure 6.11-1.

The constant k_{r3} is a sensitive parameter since if $-k_{r3}(u_{2n} - u_2) < 0$ remains in effect for a long enough time, $u_3(z,t) < 0$ (negative neuron cell density) could occur

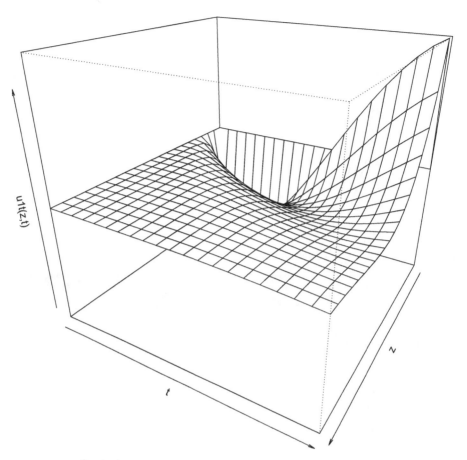

Figure 6.9-2 $\dfrac{\partial u_1(z,t)}{\partial t}$ from eqs. (3.1), 3D, ncase=2

(which is physically impossible). This example illustrates a possibly fundamental limitation of a mathematical model, that is, the model solutions can assume physically unrealistic values (e.g., negative concentrations).

In the present case, this situation could possibly be avoided by using a nonlinear, logistic rate, $-k_{r3}(u_{2n} - u_2)u_3$ so that as $u_3(z,t) \to 0$ the rate approaches zero (rather than remain negative). This extension of eq. (5.1-1) is left as an exercise.

For ncase=3, the graphical output is shown in Figures 6.12,13,14.

Figure 6.12-1 indicates $\dfrac{\partial u_1}{\partial t} < 0$ and movement toward a zero-derivative equilibrium (steady state) solution.

Figure 6.12-2 confirms Figure 6.12-1.

Figure 6.13-1 indicates $\dfrac{\partial u_2}{\partial t} < 0$ and movement toward a zero-derivative equilibrium (steady state) solution.

Case Studies

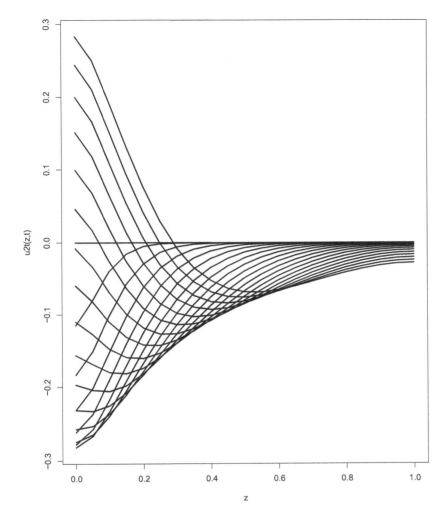

Figure 6.10-1 $\dfrac{\partial u_2(z,t)}{\partial t}$ from eqs. (3.2), 2D, ncase=2

Figure 6.13-2 confirms Figure 6.13-1.

Figure 6.14-1 indicates $\dfrac{\partial u_3}{\partial t} < 0$ and continuing decrease of u_3 with t. This decrease in $u_3(z,t)$ results from the single RHS term in eq. (5.1-1), $-k_{r3}(u_{2n} - u_2) < 0$. Figure 6.14-2 confirms Figure 6.14-1.

In summary, the reduction in O_2 concentration of the capillary blood and brain tissue, $u_1(z,t)$, $u_2(z,t)$, leads to a reduction in the neuron cell density, $u_3(z,t)$. These changes could then explain long Covid cognitive impairment.

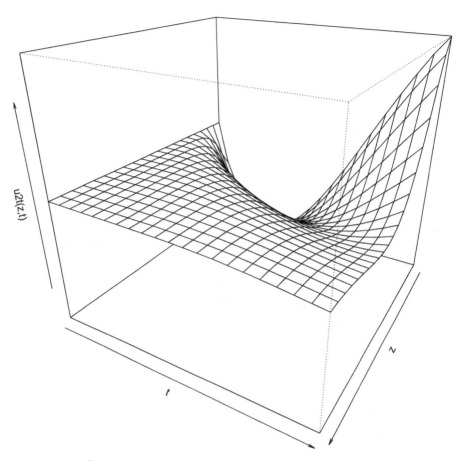

Figure 6.10-2 $\frac{\partial u_2(z,t)}{\partial t}$ from eqs. (3.2), 3D, `ncase=2`

Case Studies

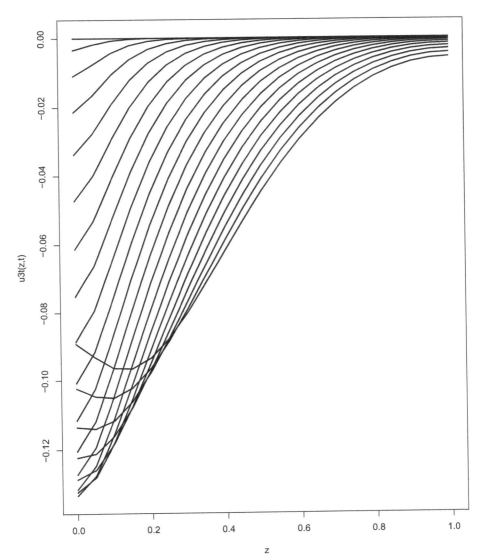

Figure 6.11-1 $\dfrac{\partial u_3(z,t)}{\partial t}$ from eqs. (5.1), 2D, `ncase=2`

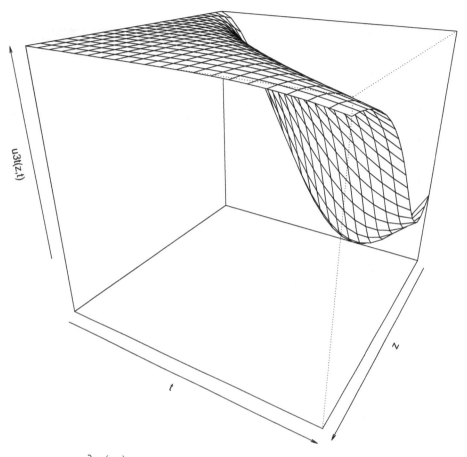

Figure 6.11-2 $\dfrac{\partial u_3(z,t)}{\partial t}$ from eqs. (5.1), 3D, `ncase=2`

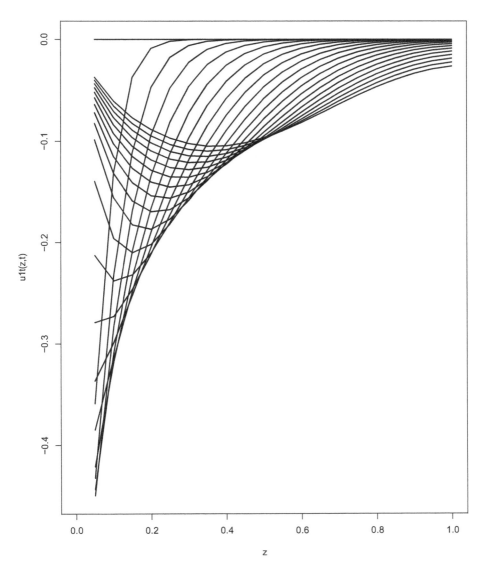

Figure 6.12-1 $\dfrac{\partial u_1(z,t)}{\partial t}$ from eqs. (3.1), 2D, ncase=3

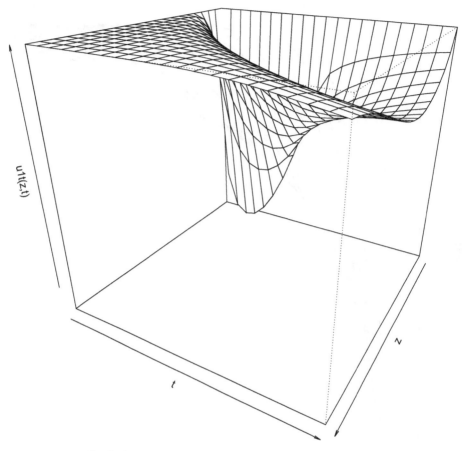

Figure 6.12-2 $\dfrac{\partial u_1(z,t)}{\partial t}$ from eqs. (3.1), 3D, ncase=3

Case Studies

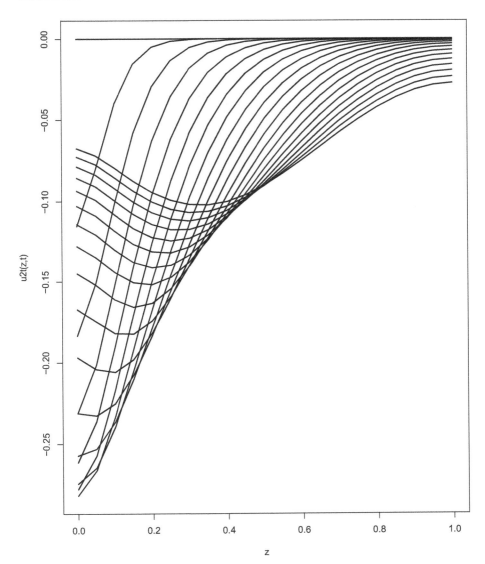

Figure 6.13-1 $\dfrac{\partial u_2(z,t)}{\partial t}$ from eqs. (3.2), 2D, `ncase=3`

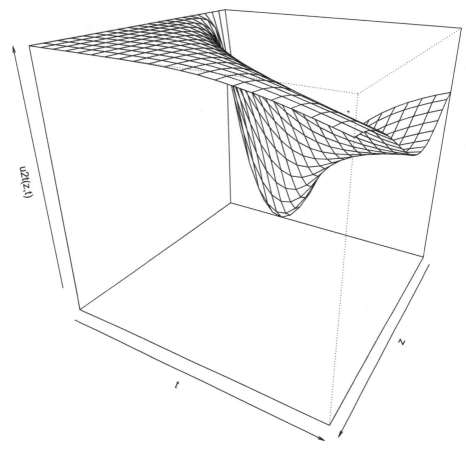

Figure 6.13-2 $\dfrac{\partial u_2(z,t)}{\partial t}$ from eqs. (3.3), 3D, `ncase=3`

6.1.3 ANALYSIS OF PDE RHS TERMS

$u_1(z,t)$, $u_2(z,t)$, $u_3(z,t)$ are determined by the time derivatives, $\dfrac{\partial u_1(z,t)}{\partial t}$, $\dfrac{\partial u_2(z,t)}{\partial t}$, $\dfrac{\partial u_3(z,t)}{\partial t}$ of eqs. (3.1-1), (3.2-1), (5.1-1). These derivatives, in turn, are defined by the RHS terms of eqs. (3.1-1), (3.2-1), (5.1-1) so that these terms provide insight into the origin of the features of the solutions. Therefore, the calculation and display of the PDE RHS terms is an important approach to the analysis of the PDE solutions. This type of analysis is demonstrated by the following programming.

Main program

For the analysis of the RHS of eq. (3.1-1), two terms are calculated and displayed.

Case Studies

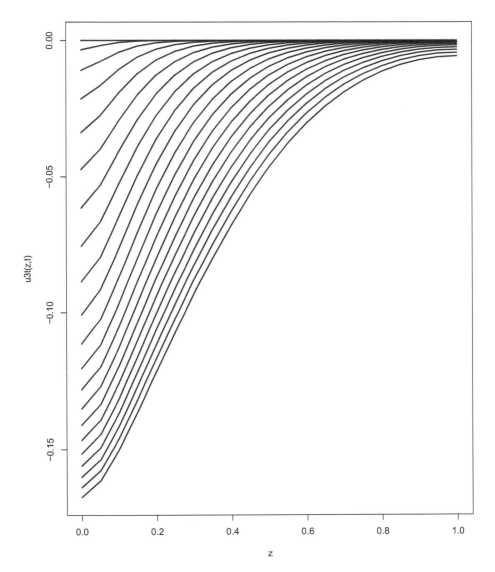

Figure 6.14-1 $\dfrac{\partial u_3(z,t)}{\partial t}$ from eqs. (5.1), 2D, ncase=3

- term11 $= -v_z \dfrac{\partial u_1}{\partial z}$.
- term12 $= -(2/r_l)k_{m1}(u_1 - u_2)$.

The naming of the terms is according to term-PDE number-term in PDE. For example, term11 is for PDE 1 (eq. (3.1-1)), term 1 (convection).

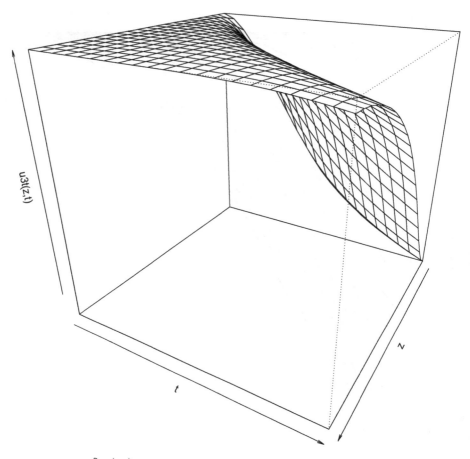

Figure 6.14-2 $\dfrac{\partial u_3(z,t)}{\partial t}$ from eqs. (5.1), 3D, ncase=3

These terms are calculated by the addition of the following code to the end of Listing 6.3.

```
#
# PDE RHS terms
#
# u1t
#
# Matrices for term11, term12
  term11=matrix(0,nrow=nz,ncol=nout);
  term12=matrix(0,nrow=nz,ncol=nout);
#
# Calculation of term11, term12
```

Case Studies

```
    for(it in 1:nout){
    for(iz in 1:nz){
      if(iz==1){term11[1,it]=-vz*(u1[2,it]-u1[1,it])/dz;}
      if(iz>1){
      term11[iz,it]=-vz*(u1[iz,it]-u1[iz-1,it])/dz;}
      term12[iz,it]=-r1*km1*(u1[iz,it]-u2[iz,it]);
    }
    }
#
# Plot term11
    par(mfrow=c(1,1));
    matplot(x=z[2:nz],y=term11[2:nz,],type="l",xlab="z",
            ylab="term11(z,t)",xlim=c(zl,zu),lty=1,main="",
            lwd=2,col="black");
    persp(z,tout,term11,theta=120,phi=25,
          xlim=c(zl,zu),ylim=c(t0,tf),
          xlab="z",ylab="t",zlab="term11(z,t)");
#
# Plot term12
    par(mfrow=c(1,1));
    matplot(x=z,y=term12,type="l",xlab="z",ylab="term12(z,t)",
            xlim=c(zl,zu),lty=1,main="",lwd=2,col="black");
    persp(z,tout,term12,theta=120,phi=25,
          xlim=c(zl,zu),ylim=c(t0,tf),
          xlab="z",ylab="t",zlab="term12");
```

Listing 6.4: Additional code for the calculation of eq. (3.1-1) RHS terms

We can note the following details about Listing 6.4.

- Matrices are defined for `term11`, `term12`.

    ```
    #
    # PDE RHS terms
    #
    # u1t
    #
    # Matrices for term11, term12
        term11=matrix(0,nrow=nz,ncol=nout);
        term12=matrix(0,nrow=nz,ncol=nout);
    ```

- `term11`, `term12` are calculated according to eq. (3.1-1).

    ```
    #
    # Calculation of term11, term12
        for(it in 1:nout){
    ```

```
     for(iz in 1:nz){
       if(iz==1){term11[1,it]=-vz*(u1[2,it]-u1[1,it])/dz;}
       if(iz>1){
       term11[iz,it]=-vz*(u1[iz,it]-u1[iz-1,it])/dz;}
       term12[iz,it]=-r1*km1*(u1[iz,it]-u2[iz,it]);
     }
     }
```

- term11, term12 are plotted in 2D and 3D.

```
  #
  # Plot term11
    par(mfrow=c(1,1));
    matplot(x=z[2:nz],y=term11[2:nz,],type="l",xlab="z",
            ylab="term11(z,t)",xlim=c(zl,zu),lty=1,main="",
            lwd=2,col="black");
    persp(z,tout,term11,theta=120,phi=25,
          xlim=c(zl,zu),ylim=c(t0,tf),
          xlab="z",ylab="t",zlab="term11(z,t)");
  #
  # Plot term12
    par(mfrow=c(1,1));
    matplot(x=z,y=term12,type="l",xlab="z",ylab="term12(z,t)",
            xlim=c(zl,zu),lty=1,main="",lwd=2,col="black");
    persp(z,tout,term12,theta=120,phi=25,
          xlim=c(zl,zu),ylim=c(t0,tf),
          xlab="z",ylab="t",zlab="term12");
```

ODE/MOL routine

The ODE/MOL routine is pde3a of Listings 5.2, 6.2.

Numerical and graphical output

The graphical output for ncase=3, (set in Listing 5.1) is in Figures 6.15,16.

Figure 6.15-1 indicates that term11 at $z = z_l$ is not included (from x=z[2:nz],y=term11[2:nz,]). Figure 6.15-2 confirms Figure 6.15-1.

Figure 6.16-1 indicates that term12 at $z = z_l$ is included (from x=z,y=term12). Figure 6.16-2 confirms Figure 6.16-1.

A similar analysis can be applied for the remaining PDE RHS terms.

- term21 = $D_2 \dfrac{\partial^2 u_2}{\partial z^2}$ (eq. (3.2-1)).
- term22 = $\dfrac{2r_l}{\left(r_u^2 - r_l^2\right)} k_{m1}(u_1 - u_2)$ (eq. (3.2-1)).
- term31 = $-k_{r3}(u_{2n} - u_2)$ (eq. (5.1-1)).

Case Studies

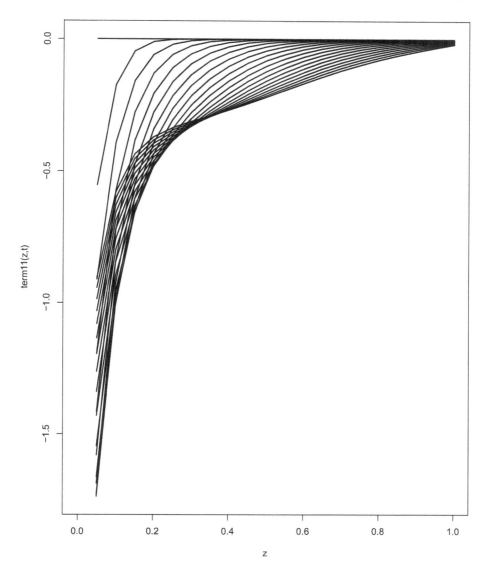

Figure 6.15-1 `term11` from Listing 6.4, 2D, `ncase=3`

The calculation and display of these terms is left as an exercise.

The computation and display of PDE RHS terms is an important methodology for evaluating the physical/chemical/biological contributions to a PDE model. For example, eq. (3.1-1) can be stated as

$$\frac{\partial u_1}{\partial t} = term_{11} + term_{12}$$

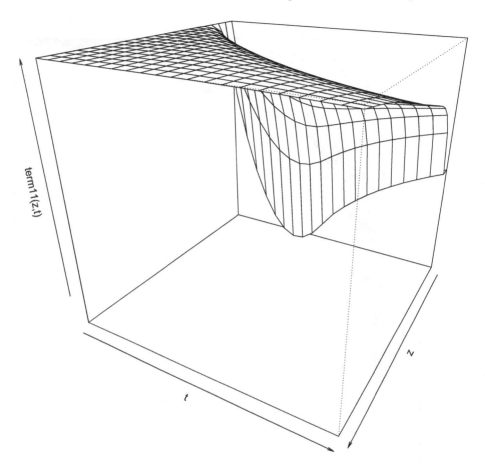

Figure 6.15-2 `term11` from Listing 6.4, 3D, `ncase=3`

so that $term_{11} + term_{12}$ defines the derivative $\dfrac{\partial u_1}{\partial t}$, which, in turn, determines through MOL numerical integration $u_1(z,t)$. By examining $term_{11}$, $term_{12}$, the relative contributions of convection and O_2 mass transfer across the BBB can be observed.

This is illustrated in Figures 6.15,16. Figure 6.15 indicates $term_{11} < 0$ while Figure 6.16 indicates $term_{12} > 0$. Furthermore, the two terms have the same general form so that they tend to sum to a small value. If they sum to zero, $\dfrac{\partial u_1}{\partial t} = 0$ corresponding to an equilibrium solution for $u_1(z,t)$. That is, convection is balanced by mass transfer. This detailed insight into the solution of eq. (3.1-1) can then be used to adjust parameters and possibly alter the form of the RHS terms (the variation of a PDE t derivative is illustrated in Figures 6.9,12. for eq. (3.1-1)).

Case Studies

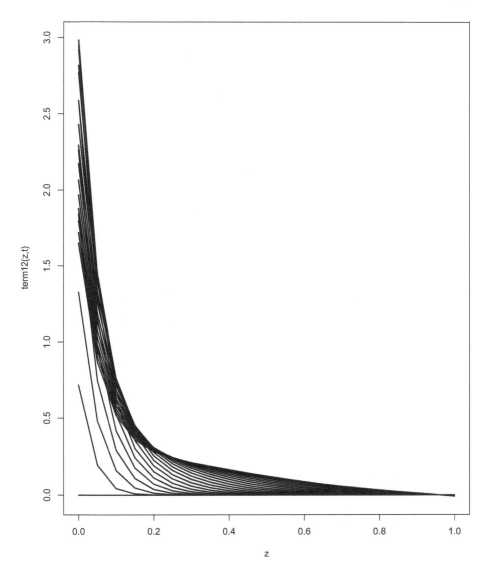

Figure 6.16-1 term12 from Listing 6.4, 2D, ncase=3

The methodology of examining the PDE RHS terms can be readily applied to all of the model PDEs once numerical solutions (PDE dependent variables) have been calculated, to provide insight into the differences of the solutions in t and z. An important special case in physical applications modeled by systems of simultaneous PDEs is when (1) certain PDEs have equilibrium solutions and (2) other PDEs have dynamic solutions. These hybrid systems are termed "stiff" and generally can be accommodated within the MOL framework.

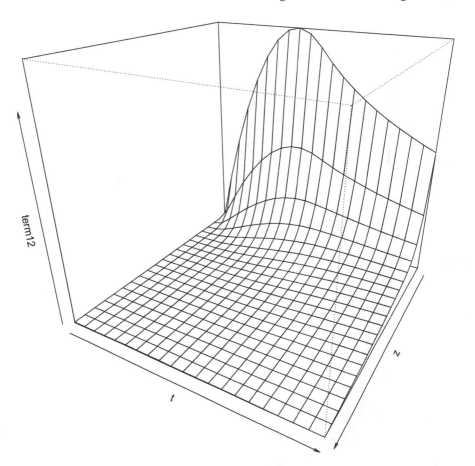

Figure 6.16-2 `term12` from Listing 6.4, 3D, `ncase=3`

6.1.4 SUMMARY AND CONCLUSIONS

Two cases of the three PDE model of eqs. (3.1), (3.2), (5.1) are developed in this chapter. For the second case, the PDE analysis is illustrated by: (1) the calculation and display of the derivatives $\dfrac{\partial u_1}{\partial t}, \dfrac{\partial u_2}{\partial t}, \dfrac{\partial u_3}{\partial t}$, and (2) the calculation and display of the PDE RHS terms. These additional calculations are based on the computed solutions $u_1(z,t)$, $u_2(z,t)$, $u_3(z,t)$ and provide additional insight into the origin of the features of the PDE solutions, particularly the approach to an equilibrium (steady state).

REFERENCE

1. Soetaert, K., J. Cash, and F. Mazzia (2012), *Solving Differential Equations in R*, Springer-Verlag, Heidelberg, Germany.

Appendix A: Introduction to PDE Analysis

The intent of this appendix is to provide an introduction to computer-based modeling based on partial differential equations (PDEs).

A.1 PDE NOTATION

If a biological/chemical/physical application is to be modeled in both space and time (spatiotemporal analysis), PDEs are an established mathematical form for the model. The notation used in the model PDEs is reviewed next.

If a model PDE defines a dependent variable, $u(z,t)$, as a function of the independent variables, z (space) and t (time), then the partial derivative of $u(z,t)$ with respect to z and t, is denoted, respectively, as $\dfrac{\partial u}{\partial z}$ and $\dfrac{\partial u}{\partial t}$.

$\dfrac{\partial u}{\partial z}$ is the rate of change of $u(z,t)$ with respect to z at constant t. $\dfrac{\partial u}{\partial t}$ is the rate of change of $u(z,t)$ with respect to t at constant z. These derivatives are first order in z and t, respectively.

To facilitate the coding (programming) of PDEs, a subscript notation for partial derivatives is often used.

$$\frac{\partial u}{\partial z} \Rightarrow u_z; \quad \frac{\partial u}{\partial t} \Rightarrow u_t$$

Derivatives of order higher than one can be considered for PDE models.

$$\frac{\partial \left(\frac{\partial u(z,t)}{\partial z}\right)}{\partial z} = \frac{\partial^2 u(z,t)}{\partial z^2} = u_{zz}$$

$$\frac{\partial \left(\frac{\partial u(z,t)}{\partial t}\right)}{\partial t} = \frac{\partial^2 u(z,t)}{\partial t^2} = u_{tt}$$

$$\frac{\partial \left(\frac{\partial u(z,t)}{\partial z}\right)}{\partial t} = \frac{\partial^2 u(z,t)}{\partial t \partial z} = u_{tz} = \frac{\partial \left(\frac{\partial u(z,t)}{\partial t}\right)}{\partial z} = \frac{\partial^2 u(z,t)}{\partial z \partial t} = u_{zt}$$

The third form of the second derivative is termed a mixed partial derivative, for which the order of differentiation in z and t is assumed to be interchangeable.

PDE spatial derivatives can be expressed in different coordinate systems. For example,

- **Cartesian coordinates**, (x,y,z): x,y,z are defined along mutually perpendicular (orthogonal) axes [1].
- **cylindrical coordinates**, (r,θ,z): r,θ,z are defined along mutually perpendicular axes. r is a radial coordinate [2]. θ is an angular coordinate. z is an axial coordinate.
- **spherical coordinates**, (r,θ,ϕ): r,θ,ϕ are defined along mutually perpendicular axes. r is a radial coordinate [3]. θ,ϕ are angular coordinates.

Other coordinate systems can be derived from general curvilinear coordinates [4]. The choice of a particular coordinate system is usually determined by the match of the boundary conditions (discussed subsequently) to the geometry of the physical system.

A.2 PDE CLASSIFICATION

PDEs can be classified geometrically which facilitates the selection of numerical methods (algorithms) for the computer solution of PDEs. The generally accepted geometrical classifications are summarized next.

A.2.1 HYPERBOLIC

Hyperbolic PDEs are of two types:

First order: If the PDE is first order in the initial value (temporal) variable t, and first order in the boundary value (spatial) variable z, it is generally termed first-order hyperbolic. As an example, the linear, first-order hyperbolic (advection) PDE equation, is

$$\frac{\partial u}{\partial t} + v_z \frac{\partial u}{\partial z} = 0 \qquad (A.1\text{-}1)$$

For a complete (well-posed) problem, one initial condition (IC) and one boundary condition (BC) are required.

$$u(z, t=0) = f(z) \qquad (A.1\text{-}2)$$

$$u(z=0, t) = g(t) \qquad (A.1\text{-}3)$$

where $f(z), g(t)$ are functions to be specified. As an example that is subsequently integrated numerically, $f(z) = 0, g(t) = 1$, for which an analytical (exact) solution is available.

If $\lambda = z - v_z t$ (Lagrangian coordinate, traveling wave), $u(z,t) = u(\lambda)$, and eqs. (A.1) can be stated in terms of λ.

$$\frac{\partial u}{\partial t} = \frac{du}{d\lambda}\frac{\partial \lambda}{\partial t} = -v_z \frac{du}{d\lambda}$$

$$\frac{\partial u}{\partial z} = \frac{du}{d\lambda}\frac{\partial \lambda}{\partial z} = \frac{du}{d\lambda}$$

Appendix A

Substitution in eq. (A.1-1) gives

$$\frac{\partial u}{\partial t} + v_z \frac{\partial u}{\partial z} = -v_z \frac{du}{d\lambda} + v_z \frac{du}{d\lambda} = 0$$

Thus, $u(\lambda)$ is a solution to eqs. (A.1) with $u(\lambda > 0) = 0$ (IC (A.1-2), $f(z) = 0$), $u(\lambda < 0) = 1$ (BC (A.1-3) $g(t) = 1$) (a unit step or Heaviside function, traveling left to right in z with velocity v_z). This special case of the advection equation is used subsequently in evaluating (testing) a numerical solution.

Second order: If the PDE is second order in the initial value (temporal) variable t, and second order in the boundary value (spatial) variable z, it is generally termed second-order hyperbolic. For example, the linear wave equation, is

$$\frac{\partial^2 u}{\partial t^2} = c^2 \frac{\partial^2 u}{\partial z^2}; \; u_{tt} = c^2 u_{zz} \tag{A.2-1}$$

where c is a wave velocity.

Equation (A.2-1) is second order in t and requires two ICs,

$$u(z,t=0) = f_1(z); \; \frac{\partial u(z, t=0)}{\partial t} = u_t(z, t=0) = f_2(z) \tag{A.2-2,3}$$

where $f_1(z), f_2(z)$ are functions to be specified.

Equation (A.2-1) is second order in z and requires two BCs, for example

$$u(z = z_l, t) = g_1(z); \; \frac{\partial u(z = z_u, t)}{\partial z} = g_2(z) \tag{A.2-4,5}$$

where z_l, z_u are lower and upper boundary values of z, respectively, and $g_1(z), g_2(z)$ are functions to be specified. Equation (A.2-4) is a Dirichlet BC[1] and eq. (A.2-5) is a Neumann BC.

A second-order hyperbolic PDE can be expressed as a system of two first-order hyperbolic PDEs. For example, $u(z,t), v(z,t)$ can be defined by two simultaneous first-order hyperbolic PDEs.

$$\frac{\partial v}{\partial z} = \frac{\partial u}{\partial t} \tag{A.3-1}$$

$$\frac{\partial v}{\partial t} = \frac{\partial u}{\partial z} \tag{A.3-2}$$

If eq. (A.3-1) is differentiated with respect to t and eq. (A.3-2) is differentiated with respect to z,

$$\frac{\partial^2 v}{\partial t \partial z} = \frac{\partial^2 u}{\partial t^2} \tag{A.3-3}$$

$$\frac{\partial^2 v}{\partial z \partial t} = \frac{\partial^2 u}{\partial z^2} \tag{A.3-4}$$

---

[1] A Dirichlet BC specifies the dependent variable, $u(z,t)$, at the boundary. A Neumann BC specifies the first-order spatial derivative of $u(z,t)$ with respect to z at the boundary. A Robin BC includes both the dependent variable and the first-order spatial derivative, frequently as a linear combination, e.g., $D \frac{\partial u(z=z_u,t)}{\partial z} + k_m u(z=z_u,t) = 0$.

If eq. (A.3-4) is subtracted from eq. (A.3-3),

$$\frac{\partial^2 v}{\partial t \partial z} - \frac{\partial^2 v}{\partial z \partial t} = \frac{\partial^2 u}{\partial t^2} - \frac{\partial^2 u}{\partial z^2}$$

or if the mixed partial derivatives are assumed equal

$$\frac{\partial^2 u}{\partial t^2} = \frac{\partial^2 u}{\partial z^2}$$

which is eq. (A.2-1) with $c = 1$. This result suggests second-order hyperbolic PDEs can be studied through associated systems of first-order PDEs.

A second-order hyperbolic PDE can also be expressed as a system of PDEs first order in t and second order in z. For example, if two variables, $u_1(z,t)$ and $u_2(z,t)$, are defined as

$$u_1(z,t) = u(z,t); \quad u_2(z,t) = \frac{\partial u}{\partial t} = \frac{\partial u_1}{\partial t}$$

eq. (A.2-1) can be written as

$$\frac{\partial u_1}{\partial t} = u_2(z,t); \quad \frac{\partial u_2}{\partial t} = \frac{\partial^2 u_1}{\partial z^2}$$

This formulation is particularly useful when numerically integrating second-order hyperbolic PDEs.

A.2.2 PARABOLIC

Parabolic PDEs are first order in an initial value variable and second order in a boundary value variable. The diffusion equation (heat conduction equation, Fick's second law, Fourier's second law) is an example.

$$\frac{\partial u}{\partial t} = D \frac{\partial^2 u}{\partial z^2} \tag{A.4-1}$$

where D is a diffusivity. Equation (A.4-1) is first order in t and requires one IC.

$$u(z, t=0) = f(z) \tag{A.4-2}$$

Equation (A.4-1) is second order in z and requires two BCs, for example

$$u(z=z_l,t) = g(t); \quad D\frac{\partial u(z=z_u,t)}{\partial z} + k_m u_z(z=z_u,t) = 0 \tag{A.4-3,4}$$

where z_l, z_u are lower and upper boundary values of z, respectively, k_m is a mass transfer coefficient, and $f(z)$, $g(t)$ are functions to be specified. Equation (A.4-3) is a Dirichlet BC and eq. (A.4-4) is a Robin BC (explained previously).

Equation (A.4-1) is based on a material (mass) balance for an incremental volume $A_c \Delta z$.

$$A_c \Delta z \frac{\partial u}{\partial t} = -A_c D \frac{\partial u}{\partial z}\bigg|_z - \left(-A_c D \frac{\partial u}{\partial z}\bigg|_{z+\Delta z}\right) \tag{A.4-5}$$

Appendix A

The diffusive flux into and out of the incremental volume is based on Fick's first law (Fourier's first law).

$$q = -D\frac{\partial u}{\partial z} \tag{A.4-6}$$

with $q > 0$ for $\frac{\partial u}{\partial z} < 0$ (diffusion in the direction of decreasing $u(z,t)$).

Division of eq. (A.4-5) by $A_c \Delta z$ and minor rearrangement gives

$$\frac{\partial u}{\partial t} = D \left(\frac{\frac{\partial u}{\partial z}\big|_{z+\Delta z} - \frac{\partial u}{\partial z}\big|_z}{\Delta z} \right) \tag{A.4-7}$$

With $\Delta z \to 0$, eq. (A.4-7) is eq. (A.4-1).

The diffusion equation can be expressed in terms of the coordinate-independent vector differential operator ∇.

$$\frac{\partial u}{\partial t} = D \nabla \cdot \nabla u = D \, \text{div grad } u \tag{A.4-8}$$

where $\nabla \cdot$ is the divergence of a vector and ∇ is the gradient of a scalar. ∇ can then be stated for a particular coordinate system. This is done in the following tables for Cartesian, cylindrical and spherical coordinates.

$\nabla \cdot$ (divergence of a vector):

Coordinate system	Components
Cartesian	$[\nabla]_x = \frac{\partial}{\partial x}$ $[\nabla]_y = \frac{\partial}{\partial y}$ $[\nabla]_z = \frac{\partial}{\partial z}$
cylindrical	$[\nabla]_r = \frac{1}{r}\frac{\partial}{\partial r}(r)$ $[\nabla]_\theta = \frac{1}{r}\frac{\partial}{\partial \theta}$ $[\nabla]_z = \frac{\partial}{\partial z}$
spherical	$[\nabla]_r = \frac{1}{r^2}\frac{\partial}{\partial r}(r^2)$ $[\nabla]_\theta = \frac{1}{r\sin\theta}\frac{\partial}{\partial \theta}(\sin\theta)$ $[\nabla]_\phi = \frac{1}{r\sin\theta}\frac{\partial}{\partial \phi}$

∇ (gradient of a scalar):

Coordinate system	Components
Cartesian	$[\nabla]_x = \dfrac{\partial}{\partial x}$ $[\nabla]_y = \dfrac{\partial}{\partial y}$ $[\nabla]_z = \dfrac{\partial}{\partial z}$
cylindrical	$[\nabla]_r = \dfrac{\partial}{\partial r}$ $[\nabla]_\theta = \dfrac{1}{r}\dfrac{\partial}{\partial \theta}$ $[\nabla]_z = \dfrac{\partial}{\partial z}$
spherical	$[\nabla]_r = \dfrac{\partial}{\partial r}$ $[\nabla]_\theta = \dfrac{1}{r}\dfrac{\partial}{\partial \theta}$ $[\nabla]_\phi = \dfrac{1}{r\sin\theta}\dfrac{\partial}{\partial \phi}$

$\nabla \cdot \nabla$ (divergence of the gradient of a scalar):

Coordinate system	Component
Cartesian	$\dfrac{\partial^2}{\partial x^2} + \dfrac{\partial^2}{\partial y^2} + \dfrac{\partial^2}{\partial z^2}$
cylindrical	$\left(\dfrac{\partial^2}{\partial r^2} + \dfrac{1}{r}\dfrac{\partial}{\partial r}\right) + \dfrac{1}{r^2}\dfrac{\partial^2}{\partial \theta^2} + \dfrac{\partial^2}{\partial z^2}$
spherical	$\left(\dfrac{\partial^2}{\partial r^2} + \dfrac{2}{r}\dfrac{\partial}{\partial r}\right) + \dfrac{1}{r^2 \sin\theta}\dfrac{\partial}{\partial \theta}\left(\sin\theta \dfrac{\partial}{\partial \theta}\right) + \dfrac{1}{r^2 \sin^2\theta}\dfrac{\partial^2}{\partial \phi^2}$

Cartesian coordinates:

With the orthonormal vectors $(\mathbf{i}_x, \mathbf{j}_y, \mathbf{k}_z)$, $\nabla \cdot \nabla$ in the RHS of eq. (A.4-8) follows as

$$\nabla \cdot \nabla = \left(\mathbf{i}\dfrac{\partial}{\partial x} + \mathbf{j}\dfrac{\partial}{\partial y} + \mathbf{k}\dfrac{\partial}{\partial z}\right) \cdot \left(\mathbf{i}\dfrac{\partial}{\partial x} + \mathbf{j}\dfrac{\partial}{\partial y} + \mathbf{k}\dfrac{\partial}{\partial z}\right) = \dfrac{\partial^2}{\partial x^2} + \dfrac{\partial^2}{\partial y^2} + \dfrac{\partial^2}{\partial z^2}$$

Appendix A

Cylindrical coordinates:

With the orthonormal vectors $(\mathbf{i}_r, \mathbf{j}_\theta, \mathbf{k}_z)$, $\nabla \cdot \nabla$ in the RHS of eq. (A.4-8) follows as

$$\nabla \cdot \nabla = \left(\mathbf{i}_r \frac{1}{r}\frac{\partial}{\partial r}(r) + \mathbf{j}_\theta \frac{1}{r}\frac{\partial}{\partial \theta} + \mathbf{k}_z \frac{\partial}{\partial z}\right) \cdot \left(\mathbf{i}_r \frac{\partial}{\partial r} + \mathbf{j}_\theta \frac{1}{r}\frac{\partial}{\partial \theta} + \mathbf{k}_z \frac{\partial}{\partial z}\right)$$

$$= \frac{1}{r}\frac{\partial}{\partial r}\left(r\frac{\partial}{\partial r}\right) + \frac{1}{r}\frac{\partial}{\partial \theta}\left(\frac{1}{r}\frac{\partial}{\partial \theta}\right) + \frac{\partial}{\partial z}\left(\frac{\partial}{\partial z}\right)$$

$$= \frac{1}{r}\left(\frac{\partial}{\partial r} + r\frac{\partial^2}{\partial r^2}\right) + \frac{1}{r^2}\frac{\partial}{\partial \theta}\frac{\partial}{\partial \theta} + \frac{\partial}{\partial z}\frac{\partial}{\partial z}$$

$$= \left(\frac{\partial^2}{\partial r^2} + \frac{1}{r}\frac{\partial}{\partial r}\right) + \frac{1}{r^2}\frac{\partial^2}{\partial \theta^2} + \frac{\partial^2}{\partial z^2}$$

Spherical coordinates:

With the orthonormal vectors $(\mathbf{i}_r, \mathbf{j}_\theta, \mathbf{k}_\phi)$, $\nabla \cdot \nabla$ in the RHS of eq. (A.4-8) follows as

$$\nabla \cdot \nabla = \left(\mathbf{i}_r \frac{1}{r^2}\frac{\partial}{\partial r}(r^2) + \mathbf{j}_\theta \frac{1}{r\sin\theta}\frac{\partial}{\partial \theta}(\sin\theta) + \mathbf{k}_\phi \frac{1}{r\sin\theta}\frac{\partial}{\partial \phi}\right)$$

$$\times \left(\mathbf{i}_r \frac{\partial}{\partial r} + \mathbf{j}_\theta \frac{1}{r}\frac{\partial}{\partial \theta} + \mathbf{k}_\phi \frac{1}{r\sin\theta}\frac{\partial}{\partial \phi}\right)$$

$$= \frac{1}{r^2}\frac{\partial}{\partial r}\left(r^2 \frac{\partial}{\partial r}\right) + \frac{1}{r\sin\theta}\frac{\partial}{\partial \theta}\left(\sin\theta \frac{1}{r}\frac{\partial}{\partial \theta}\right) + \frac{1}{r\sin\theta}\frac{\partial}{\partial \phi}\left(\frac{1}{r\sin\theta}\frac{\partial}{\partial \phi}\right)$$

$$= \left(\frac{\partial^2}{\partial r^2} + \frac{2}{r}\frac{\partial}{\partial r}\right) + \frac{1}{r^2\sin\theta}\frac{\partial}{\partial \theta}\left(\sin\theta \frac{\partial}{\partial \theta}\right) + \frac{1}{r^2\sin^2\theta}\frac{\partial^2}{\partial \phi^2}$$

Equation (A.4-8) can be extended to include a convection term.

$$\frac{\partial u}{\partial t} = (-\nabla \cdot \mathbf{v} + D\nabla \cdot \nabla)u = (-\text{div }\mathbf{v} + D\text{ div grad })u \qquad \text{(A.4-9)}$$

where $\mathbf{v}$ is a velocity vector. In the three coordinate systems,

$$\text{Cartesian}: \mathbf{v} = \mathbf{i}_x v_x + \mathbf{j}_y v_y + \mathbf{k}_z v_z$$
$$\text{cylindrical}: \mathbf{v} = \mathbf{i}_r v_r + \mathbf{j}_\theta v_\theta + \mathbf{k}_z v_z$$
$$\text{spherical}: \mathbf{v} = \mathbf{i}_r v_r + \mathbf{j}_\theta v_\theta + \mathbf{k}_\phi v_\phi$$

As a prototype model that can be used to test numerical integration algorithms for parabolic PDEs, the following IC and BCs for eq. (A.4-1) are considered subsequently.

$$u(z, t=0) = f(z) = \sin(\pi(z-z_l)/(z_u-z_l)) \qquad \text{(A.4-10)}$$
$$u(z=z_l, t) = u(z=z_u, t) = 0 \qquad \text{(A.4-11,12)}$$

Equations (A.4-11,12) define homogeneous Dirichlet BCs.

As a second case,

$$u(z, t=0) = f(z) = \cos(\pi(z-z_l)/(z_u-z_l)) \quad \text{(A.4-13)}$$

$$\frac{\partial u(z=z_l, t)}{\partial z} = \frac{\partial u(z=z_u, t)}{\partial z} = 0 \quad \text{(A.4-14,15)}$$

Equations (A.4-14,15) define homogeneous Neumann BCs.

$$D\frac{\partial u(z=z_l, t)}{\partial z} + k_m u(z=z_l, t) = 0$$

$$D\frac{\partial u(z=z_u, t)}{\partial z} + k_m u(z=z_u, t) = 0 \quad \text{(A.4-16,17)}$$

Equations (A.4-16,17) define homogeneous Robin BCs.

An exact solution to eqs. (A.4-1), (A.4-10,11,12),

$$u(z,t) = e^{-D(\pi/(z_u-z_l))^2 t} \sin(\pi(z-z_l)/(z_u-z_l)) \quad \text{(A.4-18)}$$

and an exact solution to eqs. (A.4-1), (A.4-13,14,15),

$$u(z,t) = e^{-D(\pi/(z_u-z_l))^2 t} \cos(\pi(z-z_l)/(z_u-z_l)) \quad \text{(A.4-19)}$$

are used subsequently to test numerical FD integration algorithms implemented within the method of lines (MOL) framework. An exact solution to eq. (A.4-1), (A.4-16,17) is available, but it is relatively complicated so it is not included here.

As the term diffusion equation implies (eq. (A.4-1)), parabolic PDEs smooth steep moving fronts and discontinuities, and therefore are easier to integrate numerically than hyperbolic PDEs (such as eq. (A.1-1)).

A.2.3 ELLIPTIC

Elliptic PDEs are zeroth order in an initial value variable (no derivative in t) and second order in two boundary value variables. Laplace's equation is an example.

$$\frac{\partial^2 u}{\partial y^2} + \frac{\partial^2 u}{\partial z^2} = 0 \quad \text{(A.5-1)}$$

Equation (A.5-1) is second order in y and z and requires two BCs for each. For example,

$$u(y=y_l, z) = f_1(z); \quad \frac{\partial u(y=y_u, z)}{\partial y} = f_2(z) \quad \text{(A.5-2,3)}$$

$$u(y, z=z_l) = g_1(y); \quad \frac{\partial u(y, z=z_u)}{\partial z} = g_2(z) \quad \text{(A.5-4,5)}$$

One approach to the numerical integration of elliptic PDEs is to append a derivative in an initial value variable. For example, for eq. (A.5-1),

$$\frac{\partial u}{\partial t} = \frac{\partial^2 u}{\partial y^2} + \frac{\partial^2 u}{\partial z^2} \quad \text{(A.5-6)}$$

Appendix A

that is, to effectively convert the elliptic PDE to a parabolic PDE, then integrate the parabolic equation to a steady state. For example, for eq. (A.5-6), $\frac{\partial u}{\partial t} \approx 0$, at which point the solution $u(y,z,t \to \infty)$ is the solution to the elliptic PDE.

This approach to a numerical solution of the elliptic PDE is termed the method of false transients since t is not part of the original elliptic problem, but rather, is a parameter that continues the solution of the parabolic PDE to the solution of the elliptic PDE.

A.2.4 MULTITYPE

More than one geometric form can be included in a PDE. For example, a combination of eqs. (A.1-1) and (A.4-1) gives

$$\frac{\partial u}{\partial t} = -v_z \frac{\partial u}{\partial z} + D \frac{\partial^2 u}{\partial z^2} \tag{A.6-1}$$

Equation (A.6-1) mathematically is a hyperbolic-parabolic PDE, and physically, a convection-diffusion equation.

Equation (A.6-1) is first order in t and requires one IC.

$$u(z, t = 0) = f(z) \tag{A.6-2}$$

where $f(z)$ is a function to be specified.

Equation (A.6-1) is second order in z and requires two BCs, e.g.,

$$u(z = z_l, t) = g(t); \quad \frac{\partial u(z = u, t)}{\partial z} = 0 \tag{A.6-3,4}$$

where $g(t)$ is a function to be specified.

BC (A.6-4) specifies a zero slope (derivative in z) at $z = z_u$. If the solution $u(z,t)$ is a discontinuity (step) traveling in the positive z direction ($v_z > 0$) as discussed previously, eq. (A.6-4) requires the discontinuity move across the exit boundary at $z = z_u$ with zero slope. Since this is essentially impossible, the solution displays a numerical error (artifact) such as an oscillation.

To avoid this situation of an unrealistic exit BC, the following condition can be used.

$$\frac{\partial u(z = z_u, t)}{\partial t} + v_z \frac{\partial u(z = z_u, t)}{\partial z} = 0 \tag{A.6-5}$$

Equation (A.6-5) qualifies as a BC for eq. (A.6-1) since it is of lower order in z than eq. (A.6-1), Experience has indicated that BC (A.6-5) provides a smooth exit of a discontinuity at $z = z_u$.

If a volumetric source term is added to eq. (A.6-1),

$$\frac{\partial u}{\partial t} = -v_z \frac{\partial u}{\partial z} + D \frac{\partial^2 u}{\partial z^2} + k_r u^p \tag{A.6-6}$$

eq. A.6-6 mathematically is still hyperbolic-parabolic, and physically is a convection-diffusion-reaction equation. If $p \neq 1$, eq. (A.6-6) is also nonlinear, i.e., it is not of first degree (the distinction between order and degree is noteworthy).

A.3 FIRST-ORDER SPATIAL DERIVATIVES

The implementation of eq. (A.6-6) in a series of R routines[2] is considered next, starting with the approximation of the first-order spatial derivative $\frac{\partial u}{\partial z}$.

A.3.1 FINITE DIFFERENCES

The derivative in z, $\frac{\partial u}{\partial z}$, in eqs. (A.1-1), (A.6-1), (A.6-6) can be approximated by a two-point, upwind[3] finite difference (FD).

$$\frac{\partial u}{\partial z} \approx \frac{u(z,t) - u(z-\Delta z, t)}{\Delta z} + O(\Delta z) \quad \text{(A.7-1)}$$

where Δz is the interval in z for a FD grid of n_z points

$$\Delta z = \frac{z_u - z_l}{n_z - 1} \quad \text{(A.7-2)}$$

$O(\Delta z)$ indicates that the truncation error of the FD approximation of eq. (A.7-1) is first order (first degree) in Δz.

Substitution of approximation (A.7-1) in eq. (A.1-1) gives

$$\frac{\partial u(z,t)}{\partial t} = -v_z \frac{u(z,t) - u(z-\Delta z, t)}{\Delta z} \quad \text{(A.7-3)}$$

Equation (A.7-3) is a system of ODEs[4].

$$\frac{du_i}{dt} = -v_z \frac{u_i - u_{i-1}}{\Delta z}; \quad i = 1, 2, \ldots, n_z \quad \text{(A.7-4)}$$

If IC (A.1-2) is a unit step (Heaviside function), $h(t)$,

$$h(t) = \begin{cases} 1, & t < 0 \\ 0, & t > 0 \end{cases} \quad \text{(A.7-5)}$$

the exact (analytical) solution to eq. (A.1-1) (also discussed after eq. (A.1-3)) is

$$u(z,t) = u(z - v_z t) = h(z - v_z t) \quad \text{(A.7-6)}$$

which can be used to test the MOL numerical solution, eq. (A.7-3).

---

[2] R is a quality, open-source scientific computing system that is easily downloaded from the Internet.

[3] Upwind refers to the use of $u(z - \Delta, t)$ in the FD which is upstream of $u(z,t)$ (with $v_z > 0$). As expected, $u(z - \Delta, t)$ affects $u(z,t)$ by convection. If the downwind (downstream) value $u(z + \Delta z, t)$ is used in the FD approximation, the resulting ordinary differential equation (ODE) system is unstable (unbounded $u(z,t)$ with increasing t).

[4] Equation (A.7-3) is an example of the method of lines (MOL) approximation of a PDE as a series of ODEs. The MOL is a general numerical method (algorithm) in which the spatial derivatives of a PDE are replaced with an algebraic approximation, in this case, a FD. The approximating ODEs are then integrated by a library initial value ODE integrator.

Appendix A

A main program for eqs. (a.1) and (A.7) follows.

```
#
# First order hyperbolic PDE
# (advection equation)
#
# Delete previous workspaces
  rm(list=ls(all=TRUE))
#
# Access ODE integrator
  library("deSolve");
#
# Access functions for numerical solution
  setwd("f:/Covid-19 neurological effects/appA");
  source("pde1a.R");
  source("step.R");
#
# Set case
  ncase=1;
#
# Parameters
  nz=21;
  vz=1;
  ue=1;
#
# Spatial grid in z
  zl=0;zu=1;dz=(zu-zl)/(nz-1);
  z=seq(from=zl,to=zu,by=dz);
#
# Independent variable for ODE integration
  t0=0;tf=1;nout=5;
  tout=seq(from=t0,to=tf,by=(tf-t0)/(nout-1));
#
# Initial condition (t=0)
  u0=rep(0,nz);
  ncall=0;
#
# ODE integration
  out=lsodes(y=u0,times=tout,func=pde1a,
      sparsetype="sparseint",rtol=1e-6,
      atol=1e-6,maxord=5);
  nrow(out)
  ncol(out)
#
# Arrays for plotting numerical, exact solutions
```

```
    u=matrix(0,nrow=nz,ncol=nout);
  uex=matrix(0,nrow=nz,ncol=nout);
  for(it in 1:nout){
    for(iz in 1:nz){
       u[iz,it]=out[it,iz+1];
      uex[iz,it]=step(z[iz],tout[it],vz);
    }
   u[1,it]=ue;
  }
#
# Display numerical solution
  iv=seq(from=1,to=nout,by=2);
  for(it in iv){
    cat(sprintf("\n      t      z     z-v*t       u(z,t)     uex(z,t)\n"));
    iv=seq(from=1,to=nz,by=10);
    for(iz in iv){
      lam=z[iz]-vz*tout[it];
      cat(sprintf("%6.2f%6.2f%8.2f%12.3e%12.3e\n",
          tout[it],z[iz],lam,u[iz,it],uex[iz,it]));
    }
  }
#
# Calls to ODE routine
  cat(sprintf("\n\n ncall = %5d\n\n",ncall));
#
# Plot PDE solution
#
# u
  par(mfrow=c(1,1));
  matplot(x=z[2:nz],y=u[2:nz,],type="l",xlab="z",
          ylab="u(z,t)",xlim=c(zl,zu),lty=1,main="",lwd=2,
          col="black");
  matpoints(x=z[2:nz],u[2:nz,],pch="n",lty=1,lwd=2,
          col="black");
#
# uex
  matpoints(x=z[2:nz],uex[2:nz,],type="l",lty=1,lwd=2,
          col="black");
  matpoints(x=z[2:nz],uex[2:nz,],pch="x",lty=1,lwd=2,
          col="black");
```

Listing A.1: Main program for eqs. (A.1-1,2,3), first-order hyperbolic

Appendix A

The following details about Listing A.1 can be noted.

- Previous workspaces are deleted.

  ```
  #
  # First order hyperbolic PDE
  # (advection equation)
  #
  # Delete previous workspaces
    rm(list=ls(all=TRUE))
  ```

- The R ODE integrator library deSolve is accessed [6].

  ```
  #
  # Access ODE integrator
    library("deSolve");
  #
  # Access functions for numerical solution
    setwd("f:/Covid-19 neurological effects/appA");
    source("pde1a.R");
    source("step.R");
  ```

  Then the directory with the files for the solution of eqs. (A.1-1,2,3) is designated. Note that setwd (set working directory) uses / rather than the usual \.
  The ODE/MOL routine pde1a is discussed subsequently. step is the exact (analytical) solution to eq. (A.1-1) with eqs. (A.1-2,3) specified as a Heaviside step function, $u(z,t) = h(z - v_z t)$.

- An index for a case, with ncase=1,2, is specified. This index is used in the ODE/MOL routine pde1a considered subsequently.

  ```
  #
  # Set case
    ncase=1;
  ```

- The parameters of eqs. (A.1-1,2,3) are specified

  ```
  #
  # Parameters
    nz=101;
    vz=1;
    ue=1;
  ```

  where

  - nz: number of points in the spatial grid in z.
  - vz: velocity v_z in eq. (A.1-1).

- ue: boundary value $u(z=z_l=0,t)=g(t)=u_e$ (according to BC (A.1-3)).
- A spatial grid for eq. (A.1-1) is defined with nz=101 points so that z = 0,1/100=0.01,...,1.

```
#
# Spatial grid in z
  zl=0;zu=1;dz=(zu-zl)/(nz-1);
  z=seq(from=zl,to=zu,by=dz);
```

The grid spacing is dz = Δz.
- The interval in t is defined with nout=5 output points.

```
#
# Independent variable for ODE integration
  t0=0;tf=1;nout=5;
  tout=seq(from=t0,to=tf,by=(tf-t0)/(nout-1))
```

- IC (A.1-2) is implemented, with $u_{z,t=0}=f(z)=0$.

```
#
# Initial condition (t=0)
  u0=rep(0,nz);
  ncall=0;
```

Also, the counter for the calls to pde1a is initialized.
- The system of nz=101 ODEs is integrated by the library integrator lsodes (available in deSolve, [6]). As expected, the inputs to lsodes are the ODE function, pde1a, the IC vector u0, and the vector of output values of t, tout. The length of u0 (101) informs lsodes how many ODEs are to be integrated. func,y,times are reserved names.

```
#
# ODE integration
  out=lsodes(y=u0,times=tout,func=pde1a,
      sparsetype="sparseint",rtol=1e-6,
      atol=1e-6,maxord=5);
  nrow(out)
  ncol(out)
```

nrow,ncol confirm the dimensions of out.
- $u(z,t)$ and the exact solution $h(z-v_zt)$ are placed in matrices for subsequent plotting.

```
#
# Arrays for plotting numerical, exact solutions
```

Appendix A

```
    u=matrix(0,nrow=nz,ncol=nout);
  uex=matrix(0,nrow=nz,ncol=nout);
  for(it in 1:nout){
    for(iz in 1:nz){
       u[iz,it]=out[it,iz+1];
     uex[iz,it]=step(z[iz],tout[it],vz);
    }
   u[1,it]=ue;
  }
```

Function step for the exact (analytical) solution to eq. (A.1-1), listed next, follows from eq. (A.7-6) and the discussion after eq. (A.1-3) pertaining to $\lambda = z - v_z t$ (with $h(\lambda = 0) = 0.5$).

```
  step=function(z,t,vz){
#
# Function step computes the exact (analytical)
# for the linear advection equation with step BC
#
# lambda = z-vz*t
  lam=z-vz*t
#
# Step function
  if(lam<0){step=1;}
  if(lam==0){step=0.5;}
  if(lam>0){step=0;}
#
# Return function
  return(c(step));
  }
```

Listing A.2: $h(z - v_z t)$ for eqs. (A.1)

c is the vector operator in R, i.e., step is returned as a vector to the main program of Listing A.1.
- The numerical values of $u(z,t)$ returned by lsodes and the exact solution are displayed. Every second value in t and every tenth value in z appear from by=2,10.

```
#
# Display numerical solution
  iv=seq(from=1,to=nout,by=2);
  for(it in iv){
    cat(sprintf("\n    t      z     z-v*t    u(z,t)    uex(z,t)\n"));
    iv=seq(from=1,to=nz,by=10);
    for(iz in iv){
```

```
      lam=z[iz]-vz*tout[it];
      cat(sprintf("%6.2f%6.2f%8.2f%12.3e%12.3e\n",
          tout[it],z[iz],lam,u[iz,it],uex[iz,it]));
    }
  }
```

- The number of calls to pde1a is displayed at the end of the solution.

```
#
# Calls to ODE routine
  cat(sprintf("\n\n  ncall = %5d\n\n",ncall));
```

- The numerical solution is plotted as lines and superimposed points with the letter n.

```
#
# Plot PDE solution
#
# u
  par(mfrow=c(1,1));
  matplot(x=z[2:nz],y=u[2:nz,],type="l",xlab="z",
          ylab="u(z,t)",xlim=c(zl,zu),lty=1,main="",lwd=2,
          col="black");
  matpoints(x=z[2:nz],u[2:nz,],pch="n",lty=1,lwd=2,
          col="black");
```

- The exact solution is plotted as lines and superimposed points with the letter x.

```
#
# uex
  matpoints(x=z[2:nz],uex[2:nz,],type="l",lty=1,lwd=2,
          col="black");
  matpoints(x=z[2:nz],uex[2:nz,],pch="x",lty=1,lwd=2,
          col="black");
```

The discontinuity at $z = z_l$ of eqs (A.7-5,6) is not plotted ([2:nz,]) to improve the appearance of the plots.

This completes the discussion of the main program of Listing A.1. The ODE/MOL routine, pde1a, called by lsodes in the main program is considered next.

```
  pde1a=function(t,u,parm){
#
# Function pde1a computes the t derivative
# of u(z,t)
```

Appendix A

```
#
# BC, z=zl
  u[1]=ue;
#
# PDE
  ut=rep(0,nz);
  if(ncase==1){
    for(iz in 1:nz){
      if(iz==1){ut[iz]=0;}
      if(iz>1){
        ut[iz]=-vz*(u[iz]-u[iz-1])/dz;}
    }
  }
  if(ncase==2){
    for(iz in 1:(nz-1)){
      if(iz==1){ut[iz]=0;}
      if(iz>1){
        ut[iz]=-vz*(u[iz+1]-u[iz-1])/(2*dz);}
    }
        ut[nz]=-vz*(3*u[nz]-4*u[nz-1]+u[nz-2])/(2*dz);
  }
#
# Increment calls to pde1a
  ncall <<- ncall+1;
#
# Return derivative vector
  return(list(c(ut)));
  }
```

Listing A.3: ODE/MOL routine pde1a for eq. (A.1)

The following details about Listing A.3 can be noted.

- The function is defined.

  ```
    pde1a=function(t,u,parm){
  #
  # Function pde1a computes the t derivative
  # of u1(z,t)
  ```

  t is the current value of t in eqs. (A.1). u is the 101-vector of ODE/PDE-dependent variables. parm is an argument to pass parameters to pde1a (unused, but required in the argument list). The arguments must be listed in the order stated to properly interface with lsodes called in the main program of Listing A.1. The derivative vector of the LHS of eq. (A.1-1) is calculated and returned to lsodes as explained subsequently.

- BC (A.1-3) is programmed (with $f(z) = ue$).

```
#
# BC, z=zl
  u[1]=ue;
```
- Equation (A.1-1) is programmed for ncase=1.

```
#
# PDE
  ut=rep(0,nz);
  if(ncase==1){
    for(iz in 1:nz){
      if(iz==1){ut[iz]=0;}
      if(iz>1){
        ut[iz]=-vz*(u[iz]-u[iz-1])/dz;}
    }
  }
```

The derivative $\dfrac{\partial u}{\partial z}$ in eq. (A.1-1) is approximated with a two point upwind finite difference (FD) as in eqs. (A.7).

For $z = z_l$, BC (A.1-3) sets the value of $u(z = z_l, t)$ and therefore the derivative is set to zero (if(iz==1)u1t[iz]=0;) to ensure the boundary value is maintained.

The correspondence of the programming with eqs. (A.7) is an important feature of the MOL. The principal disadvantage of the two point upwind FD, numerical diffusion which is an error (artifact) in the numerical solution, is demonstrated by the graphical output in Figure A.1-1 produced from the main program of Listing A.1 (discussed next).

- Equation (A.1-1) is programmed for ncase=2.

```
if(ncase==2){
  for(iz in 1:(nz-1)){
    if(iz==1){ut[iz]=0;}
    if(iz>1){
      ut[iz]=-vz*(u[iz+1]-u[iz-1])/(2*dz);}
  }
  ut[nz]=-vz*(3*u[nz]-4*u[nz-1]+u[nz-2])/(2*dz);
}
```

The derivative $\dfrac{\partial u}{\partial z}$ in eq. (A.1-1) is approximated with a two point centered finite difference (FD).

$$\frac{\partial u}{\partial z} \approx \frac{u(z+\Delta,t) - u(z-\Delta z,t)}{2\Delta z} + O(\Delta z^2) \qquad (A.8\text{-}1)$$

$O(\Delta z^2)$ indicates that the truncation error of the FD approximation of eq. (A.8-1) is second order in Δz.

Appendix A

Substitution of approximation (A.8-1) in eq. (A.1-1) gives

$$\frac{\partial u(z,t)}{\partial t} = -v_z \frac{u(z+\Delta z,t) - u(z-\Delta z,t)}{2\Delta z} \quad \text{(A.8-2)}$$

or in terms of the index i on the spatial grid in z,

$$\frac{du_i}{dt} = -v_z \frac{u_{i+1} - u_{i-1}}{2\Delta z}; \ i = 1, 2, ..., n_z \quad \text{(A.8-3)}$$

For the right boundary ($z = z_u = 1$), FD (A.8-1) introduces a fictitious value $u(z_u + \Delta z, t)$. To circumvent this problem, a three point concentred FD approximation for $\dfrac{\partial u(z_u,t)}{\partial z}$ is used.

$$\frac{\partial u(z_u,t)}{\partial z} \approx \frac{3u(z_u,t) - 4u(z_u - \Delta z,t) + u(z_u - 2\Delta z,t)}{2\Delta z} + O(\Delta z^2) \quad \text{(A.8-4)}$$

The centered FD approximation of eq. (A.8-1) introduces a second form of error, numerical oscillation, as demonstrated by the graphical output in Figure A.1-2 discussed subsequently.

- The counter for the calls to pde1a is incremented and returned to the main program of Listing A.1 by <<-.

```
#
# Increment calls to pde1a
  ncall <<- ncall+1;
```

- The vector ut is returned as a list as required by lsodes. c is the R vector utility. The final } concludes pde1a.

```
#
# Return derivative vector
  return(list(c(ut)));
  }
```

The numerical output from the routines of Listings A.1, A.2, A.3 follows. For ncase=1 the numerical output is in Table A.1.

[1] 5

[1] 102

t	z	z-v*t	u(z,t)	uex(z,t)
0.00	0.00	0.00	1.000e+00	5.000e-01
0.00	0.10	0.10	0.000e+00	0.000e+00
0.00	0.20	0.20	0.000e+00	0.000e+00
0.00	0.30	0.30	0.000e+00	0.000e+00

0.00	0.40	0.40	0.000e+00	0.000e+00
0.00	0.50	0.50	0.000e+00	0.000e+00
0.00	0.60	0.60	0.000e+00	0.000e+00
0.00	0.70	0.70	0.000e+00	0.000e+00
0.00	0.80	0.80	0.000e+00	0.000e+00
0.00	0.90	0.90	0.000e+00	0.000e+00
0.00	1.00	1.00	0.000e+00	0.000e+00

t	z	z-v*t	u(z,t)	uex(z,t)
0.50	0.00	-0.50	1.000e+00	1.000e+00
0.50	0.10	-0.40	1.000e+00	1.000e+00
0.50	0.20	-0.30	1.000e+00	1.000e+00
0.50	0.30	-0.20	9.991e-01	1.000e+00
0.50	0.40	-0.10	9.354e-01	1.000e+00
0.50	0.50	0.00	5.188e-01	5.000e-01
0.50	0.60	0.10	9.227e-02	0.000e+00
0.50	0.70	0.20	4.334e-03	0.000e+00
0.50	0.80	0.30	5.701e-05	0.000e+00
0.50	0.90	0.40	2.386e-07	0.000e+00
0.50	1.00	0.50	3.641e-10	0.000e+00

t	z	z-v*t	u(z,t)	uex(z,t)
1.00	0.00	-1.00	1.000e+00	1.000e+00
1.00	0.10	-0.90	1.000e+00	1.000e+00
1.00	0.20	-0.80	1.000e+00	1.000e+00
1.00	0.30	-0.70	1.000e+00	1.000e+00
1.00	0.40	-0.60	1.000e+00	1.000e+00
1.00	0.50	-0.50	1.000e+00	1.000e+00
1.00	0.60	-0.40	1.000e+00	1.000e+00
1.00	0.70	-0.30	9.993e-01	1.000e+00
1.00	0.80	-0.20	9.826e-01	1.000e+00
1.00	0.90	-0.10	8.536e-01	1.000e+00
1.00	1.00	0.00	5.133e-01	5.000e-01

```
ncall =    307
```

Table A.1: Numerical output from Listings A.1, A.2, A.3, ncase=1 eqs. (A.7)

The following details about this output can be noted.

- 5 t output points as the first dimension of the solution matrix out from lsodes as programmed in the main program of Listing A.1 (with nout=5).
- The solution matrix out returned by lsodes has 102 elements as a second dimension. The first element is the value of t. Elements 2 to 102 are $u(z,t)$ from eqs. (A.7) (for each of the 101 output points).

Appendix A

- The solution is displayed for t=0,1/4=0.25,...,1 as programmed in Listing A.1 (every second value of t is displayed as explained previously).
- The solution is displayed for z=0,1/100=0.01,...,1 as programmed in Listing A.1 (every tenth value of z is displayed as explained previously).
- IC (A.7-5) is confirmed ($t = 0$). Note in particular the IC at $z - v_z t = 0$.
- The BC $u(z = 0, t) = u_e = 1$ as programmed in Listing A.3 is confirmed.
- The computational effort as indicated by ncall = 307 is modest so that lsodes computed the solution to eqs. (A.7) efficiently.

The graphical output is in Figures A.1-1.

Comparison of the numerical solution $u(z,t)$ (plotted with n) and exact solution $h(z - v_z t)$ (plotted with x) indicates the numerical diffusion of the two point upwind FD approximation of the spatial derivative $\frac{\partial u}{\partial z}$. Generally, the two solutions (numerical and exact) are fronts moving left to right (as expected with $v_z > 0$).

For ncase=2, the numerical output is in Table A.2.

[1] 5

[1] 102

t	z	z-v*t	u(z,t)	uex(z,t)
0.00	0.00	0.00	1.000e+00	5.000e-01
0.00	0.10	0.10	0.000e+00	0.000e+00
0.00	0.20	0.20	0.000e+00	0.000e+00
0.00	0.30	0.30	0.000e+00	0.000e+00
0.00	0.40	0.40	0.000e+00	0.000e+00
0.00	0.50	0.50	0.000e+00	0.000e+00
0.00	0.60	0.60	0.000e+00	0.000e+00
0.00	0.70	0.70	0.000e+00	0.000e+00
0.00	0.80	0.80	0.000e+00	0.000e+00
0.00	0.90	0.90	0.000e+00	0.000e+00
0.00	1.00	1.00	0.000e+00	0.000e+00

t	z	z-v*t	u(z,t)	uex(z,t)
0.50	0.00	-0.50	1.000e+00	1.000e+00
0.50	0.10	-0.40	1.002e+00	1.000e+00
0.50	0.20	-0.30	1.009e+00	1.000e+00
0.50	0.30	-0.20	9.115e-01	1.000e+00
0.50	0.40	-0.10	9.638e-01	1.000e+00
0.50	0.50	0.00	3.426e-01	5.000e-01
0.50	0.60	0.10	1.753e-03	0.000e+00
0.50	0.70	0.20	6.549e-07	0.000e+00
0.50	0.80	0.30	3.609e-11	0.000e+00
0.50	0.90	0.40	4.309e-16	0.000e+00
0.50	1.00	0.50	4.286e-22	0.000e+00

t	z	z-v*t	u(z,t)	uex(z,t)
1.00	0.00	-1.00	1.000e+00	1.000e+00
1.00	0.10	-0.90	1.006e+00	1.000e+00
1.00	0.20	-0.80	1.010e+00	1.000e+00
1.00	0.30	-0.70	1.002e+00	1.000e+00
1.00	0.40	-0.60	1.017e+00	1.000e+00
1.00	0.50	-0.50	1.044e+00	1.000e+00
1.00	0.60	-0.40	9.336e-01	1.000e+00
1.00	0.70	-0.30	1.066e+00	1.000e+00
1.00	0.80	-0.20	1.133e+00	1.000e+00
1.00	0.90	-0.10	1.211e+00	1.000e+00
1.00	1.00	0.00	3.394e-01	5.000e-01

ncall =   550

Table A.2: Numerical output from Listings A.1, A.2, A.3, ncase=2, eqs. (A.8)

The following details about this output can be noted (some of the discussion for ncase=1 is repeated to provide a self contained explanation)

- 5 t output points as the first dimension of the solution matrix out from lsodes as programmed in the main program of Listing A.1 (with nout=5).
- The solution matrix out returned by lsodes has 102 elements as a second dimension. The first element is the value of t. Elements 2 to 102 are $u(z,t)$ from eqs. (A.8) (for each of the 101 output points).
- The solution is displayed for t=0,1/4=0.25,...,1 as programmed in Listing A.1 (every second value of t is displayed as explained previously).
- The solution is displayed for z=0,1/100=0.01,...,1 as programmed in Listing A.1 (every tenth value of z is displayed as explained previously).
- IC (A.7-5) is confirmed ($t = 0$). Note, in particular, the IC at $z - v_z t = 0$.
- The BC $u(z = 0, t) = u_e = 1$ as programmed in Listing A.3 is confirmed.
- The computational effort as indicated by ncall = 550 is modest so that lsodes computed the solution to eqs. (A.8) efficiently.

The graphical output is in Figures A.1-2. Comparison of the numerical solution $u(z,t)$ (plotted with n) and exact solution $h(z - v_z t)$ (plotted with x) indicates the numerical oscillation of the two point centered FD approximation of the spatial derivative $\frac{\partial u}{\partial z}$. (the numerical oscillation invalidates the numerical solution). Generally, centered FD approximations should not be applied to hyperbolic (strongly convective) PDEs.

An improvement in the numerical integration of the advection equation, eq. (A.1-1) (reduced numerical diffusion, elimination of numerical oscillation), can be achieved by using flux limiter approximations for the derivative $\frac{\partial u}{\partial z}$ ([5], pp. 37–43).

Appendix A

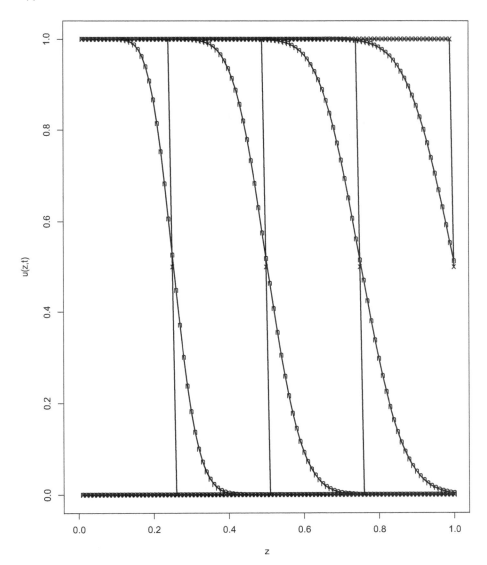

Figure A.1-1 $u(z,t)$ from eqs. (A.7)

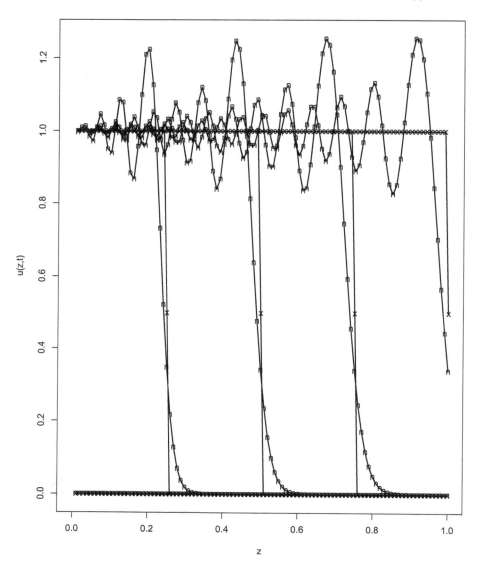

Figure A.1-2 $u(z,t)$ from eqs. (A.8)

A.3.2 FLUX LIMITERS

Consideration is now given to two flux limiters: van leer and smart. Small extensions to the main program of Listing A.1 are added, i.e., (1) files for the two flux limiters are specified and (2) an index, ncase=1,2, is added and displayed for the limiters.

```
    .
    .
    .
#
# Access functions for numerical solution
  setwd("f:/Covid-19 neurological effects/appA");
  source("pde1b.R");
  source("vanl.R");
  source("smart.R");
  source("step.R");
#
# Specify case
  ncase=1;
    .
    .
    .
#
# Display numerical solution
  iv=seq(from=1,to=nout,by=2);
   cat(sprintf("\n ncase = %2d"));
  for(it in iv){
    cat(sprintf("\n    t      z      z-v*t     u(z,t)      uex(z,t)\n"));
    iv=seq(from=1,to=nz,by=5);
    for(iz in iv){
      lam=z[iz]-vz*tout[it];
      cat(sprintf("%6.2f%6.2f%8.2f%12.3e%12.3e\n",
          tout[it],z[iz],lam,u[iz,it],uex[iz,it]));
    }
  }
    .
    .
    .
```

Listing A.4: Additions to the main program of Listing A.1 for eqs. (A.1-1,2,3)

ncase is used in the ODE/MOL routine, pde1b, to call a particular flux limiter.

```
  pde1b=function(t,u,parm){
#
# Function pde1b computes the t derivative
```

```
# of u(z,t)
#
# BC, z=zl
  u[1]=ue;
#
# PDE
  if(ncase==1){uz= van1(zl,zu,nz,u,vz);}
  if(ncase==2){uz=smart(zl,zu,nz,u,vz);}
  ut=rep(0,nz);
  for(iz in 1:nz){
    if(iz==1){ut[iz]=0;}
    if(iz>1){
      ut[iz]=-vz*uz[iz];}
  }
#
# Increment calls to pde1b
  ncall <<- ncall+1;
#
# Return derivative vector
  return(list(c(ut)));
  }
```

Listing A.5: ODE/MOL routine pde1b for eqs. (A.1-1,2,3)

The following details about Listing A.5 can be noted. (with some repetition of the discussion of Listing A.3).

- The function is defined.

    ```
    pde1b=function(t,u,parm){
    #
    # Function pde1b computes the t derivative
    # of u1(z,t)
    ```

    t is the current value of t in eqs. (A.1). u is the 101-vector of ODE/PDE dependent variables. parm is an argument to pass parameters to pde1b (unused, but required in the argument list). The arguments must be listed in the order stated to properly interface with lsodes called in the main program of Listing A.1. The derivative vector of the LHS of eq. (A.1-1) is calculated and returned to lsodes as explained subsequently.
- BC (A.1-3) is programmed (with $f(z) = ue$).

    ```
    #
    # BC, z=zl
      u[1]=ue;
    ```

Appendix A

- Equation (A.1-1) is programmed for ncase=1,2.

```
#
# PDE
  if(ncase==1){uz= vanl(zl,zu,nz,u,vz);}
  if(ncase==2){uz=smart(zl,zu,nz,u,vz);}
  ut=rep(0,nz);
  for(iz in 1:nz){
    if(iz==1){ut[iz]=0;}
    if(iz>1){
      ut[iz]=-vz*uz[iz];}
  }
```

The derivative $\frac{\partial u}{\partial z}$ in eq. (A.1-1) is approximated with a flux limiter selected by ncase. The code (programming) in functions vanl, smart is discussed subsequently.

For $z = z_l$, BC (A.1-3) sets the value of $u(z=z_l,t)$ and therefore the derivative is set to zero (if(iz==1)u1t[iz]=0;) to ensure the boundary value is maintained.

- The counter for the calls to pde1b is incremented and returned to the main program of Listing A.1 by <<-.

```
#
# Increment calls to pde1b
  ncall <<- ncall+1;
```

- The vector ut is returned as a list as required by lsodes. c is the R vector utility. The final } concludes pde1b.

```
#
# Return derivative vector
  return(list(c(ut)));
  }
```

Function vanl for the van Leer flux limiter is listed next

```
vanl=function(zl,zu,n,u,v) {
#
# Function vanl computes the van Leer flux limiter
# approximation of a first derivative
#
# Declare arrays
  uz=rep(0,n);
 phi=rep(0,n);
   r=rep(0,n);
#
```

```
# Grid spacing
  dz=(zu-zl)/(n-1);
#
# Tolerance for limiter switching
  delta=1.0e-05;
#
# Positive v
  if(v >= 0){
    for(i in 3:(n-1)){
      if(abs(u[i]-u[i-1])<delta)
         phi[i]=0
      else{
        r[i]=(u[i+1]-u[i])/(u[i]-u[i-1])
        if(r[i]<0)
           phi[i]=0
        else
        phi[i]=max(0,min(2*r[i],min(0.5*(1.0+r[i]),2)))}
      if(abs(u[i-1]-u[i-2])<delta)
         phi[i-1]=0
      else{
        r[i-1]=(u[i]-u[i-1])/(u[i-1]-u[i-2])
        if(r[i-1]<0)
           phi[i-1]=0
        else
           phi[i-1]=max(0,min(2*r[i-1],min(0.5*(1.0+r[i-1]),2)))}
      flux2=u[i  ]+(u[i  ]-u[i-1])*phi[i  ]/2;
      flux1=u[i-1]+(u[i-1]-u[i-2])*phi[i-1]/2;
      uz[i]=(flux2-flux1)/dz;
    }
    uz[1]=(-u[1]+u[2])/dz;
    uz[2]=(-u[1]+u[2])/dz;
    uz[n]=(u[n]-u[n-1])/dz;
  }
#
# Negative v
  if(v < 0){
    for(i in 2:(n-2)){
      if(abs(u[i]-u[i+1])<delta)
         phi[i]=0
      else{
        r[i]=(u[i-1]-u[i])/(u[i]-u[i+1])
        if(r[i]<0)
           phi[i]=0
        else
```

Appendix A

```
          phi[i]=max(0,min(2*r[i],min(0.5*(1.0+r[i]),2)))}
      if(abs(u[i+1]-u[i+2])<delta)
        phi[i+1]=0
      else{
        r[i+1]=(u[i]-u[i+1])/(u[i+1]-u[i+2])
        if(r[i+1]<0)
          phi[i+1]=0
        else
          phi[i+1]=max(0,min(2*r[i+1],min(0.5*(1.0+r[i+1]),2)))}
      flux2=u[i  ]+(u[i  ]-u[i+1])*phi[i  ]/2;
      flux1=u[i+1]+(u[i+1]-u[i+2])*phi[i+1]/2;
      uz[i]=-(flux2-flux1)/dz;
    }
    uz[1]=(-u[1]+u[2])/dz;
    uz[n-1]=(-u[n-1]+u[n])/dz;
    uz[n]  =(-u[n-1]+u[n])/dz;
  }
#
# All points concluded (z=zl,...,z=zu)
  return(c(uz));
}
```

Listing A.6: Function van1

The following details about van1 can be noted.

- The function is defined.

  ```
    van1=function(zl,zu,n,u,v) {
  #
  # Function van1 computes the van Leer flux limiter
  # approximation of a first derivative
  ```

  zl,zu are the lower and upper boundary values of the spatial grid in z. n is the number of grid points in z. u is the vector of PDE-dependent variable values to be differentiated numerically. v is the convective velocity (only the sign is used to determine the direction of the convection).

- Arrays (vectors) are declared for subsequent use.

  ```
  #
  # Declare arrays
    uz=rep(0,n);
   phi=rep(0,n);
     r=rep(0,n);
  ```

- The grid spacing is calculated.

```
#
# Grid spacing
  dz=(zu-zl)/(n-1);
```

- The tolerance (threshold) for switching between sections of the flux limiter is specified.

```
#
# Tolerance for limiter switching
  delta=1.0e-05;
```

delta can be varied to determine the sensitivity of the solution to its value.

- For positive velocity (in eq. (A.1-1)), the limiting function phi is calculated.

```
#
# Positive v
  if(v >= 0){
    for(i in 3:(n-1)){
      if(abs(u[i]-u[i-1])<delta)
        phi[i]=0
      else{
        r[i]=(u[i+1]-u[i])/(u[i]-u[i-1])
        if(r[i]<0)
          phi[i]=0
        else
          phi[i]=max(0,min(2*r[i],min(0.5*(1.0+r[i]),2)))}
      if(abs(u[i-1]-u[i-2])<delta)
        phi[i-1]=0
      else{
        r[i-1]=(u[i]-u[i-1])/(u[i-1]-u[i-2])
        if(r[i-1]<0)
          phi[i-1]=0
        else
          phi[i-1]=max(0,min(2*r[i-1],
                      min(0.5*(1.0+r[i-1]),2)))}
      flux2=u[i  ]+(u[i  ]-u[i-1])*phi[i  ]/2;
      flux1=u[i-1]+(u[i-1]-u[i-2])*phi[i-1]/2;
      uz[i]=(flux2-flux1)/dz;
    }
```

phi is a nonlinear function of u[i+1],u[i],u[i-1],u[i-2] (the two point upwind and centered FDs considered previously are linear in $u(z,t)$). max,min are utilities in the basic R.

Appendix A

At the end, the limiting function phi is used to calculate the fluxes flux1, flux2 and derivative in z, uz, for grid points 3:(n-1). Mathematical and graphical details for the limiting function are available in [5], pp. 41–43.
- The derivative in z is computed for the boundary points 1,2,n.

```
        uz[1]=(-u[1]+u[2])/dz;
        uz[2]=(-u[1]+u[2])/dz;
        uz[n]=(u[n]-u[n-1])/dz;
   }
```

This completes the calculation of the $\frac{\partial u}{\partial z}$ approximation for v>0.

The code for v<0 is similar with the values u[i-1],u[i-2] replaced by u[i+1],u[i+2].

The performance of the van Leer limiter is demonstrated by execution of the R routines in Listings A.1-5 with ncase=1. The numerical output is given in Table A.3.

[1] 5

[1] 102

ncase =   1

t	z	z-v*t	u(z,t)	uex(z,t)
0.00	0.00	0.00	1.000e+00	5.000e-01
0.00	0.10	0.10	0.000e+00	0.000e+00
0.00	0.20	0.20	0.000e+00	0.000e+00
0.00	0.30	0.30	0.000e+00	0.000e+00
0.00	0.40	0.40	0.000e+00	0.000e+00
0.00	0.50	0.50	0.000e+00	0.000e+00
0.00	0.60	0.60	0.000e+00	0.000e+00
0.00	0.70	0.70	0.000e+00	0.000e+00
0.00	0.80	0.80	0.000e+00	0.000e+00
0.00	0.90	0.90	0.000e+00	0.000e+00
0.00	1.00	1.00	0.000e+00	0.000e+00

t	z	z-v*t	u(z,t)	uex(z,t)
0.50	0.00	-0.50	1.000e+00	1.000e+00
0.50	0.10	-0.40	1.000e+00	1.000e+00
0.50	0.20	-0.30	1.000e+00	1.000e+00
0.50	0.30	-0.20	1.000e+00	1.000e+00
0.50	0.40	-0.10	1.000e+00	1.000e+00
0.50	0.50	0.00	4.418e-01	5.000e-01
0.50	0.60	0.10	-3.977e-07	0.000e+00
0.50	0.70	0.20	-1.773e-08	0.000e+00

```
0.50    0.80      0.30  -3.050e-10    0.000e+00
0.50    0.90      0.40  -1.527e-12    0.000e+00
0.50    1.00      0.50  -2.378e-15    0.000e+00

   t       z     z-v*t     u(z,t)       uex(z,t)
1.00    0.00     -1.00    1.000e+00    1.000e+00
1.00    0.10     -0.90    1.000e+00    1.000e+00
1.00    0.20     -0.80    1.000e+00    1.000e+00
1.00    0.30     -0.70    1.000e+00    1.000e+00
1.00    0.40     -0.60    1.000e+00    1.000e+00
1.00    0.50     -0.50    1.000e+00    1.000e+00
1.00    0.60     -0.40    1.000e+00    1.000e+00
1.00    0.70     -0.30    1.000e+00    1.000e+00
1.00    0.80     -0.20    1.000e+00    1.000e+00
1.00    0.90     -0.10    9.999e-01    1.000e+00
1.00    1.00      0.00    3.566e-01    5.000e-01

ncall =   4919
```

Table A.3: Numerical output from Listings A.1-5, ncase=1, van Leer limiter

The following details about Table A.3 can be noted.

- The dimensions of the solution matrix out returned by lsodes in the main program of Listing A.1 are again 5×102 as explained previously for Tables A.1, A.2.
- The case is confirmed (as set in the main program of Listing A.1).

  ncase = 1

- The solution is displayed for z=0,1/100=0.01,...,1 as programmed in Listing A.1 (every tenth value of z is displayed as explained previously).
- IC (A.7-5) is confirmed ($t = 0$). Note, in particular, the IC at $z - v_z t = 0$.
- The BC $u(z = 0, t) = u_e = 1$ as programmed in Listing A.5 is confirmed.
- The computational effort as indicated by ncall = 4919 is higher than for the FD output of Tables A.1, A.2 due to the computations in the van Leer limiter of Listing A.6.

The graphical output is in Figure A.2-1. Comparison of the numerical solution $u(z,t)$ (plotted with n) and exact solution $h(z - v_z t)$ (plotted with x) indicates small numerical diffusion and no numerical oscillation, i.e., a major improvement over the FD solutions in Figures A.1-1, A.1-2.

The exact solution $h(z - v_z t)$ is impossible to reproduce numerically since for $z - v_z t = 0$ the solution is undefined (infinite slope). With this in mind, the performance of the van Leer limiter is quite impressive.

Function smart for the smart flux limiter is listed next (ncase=2 in the main program of Listings A.1, A.4).

Appendix A

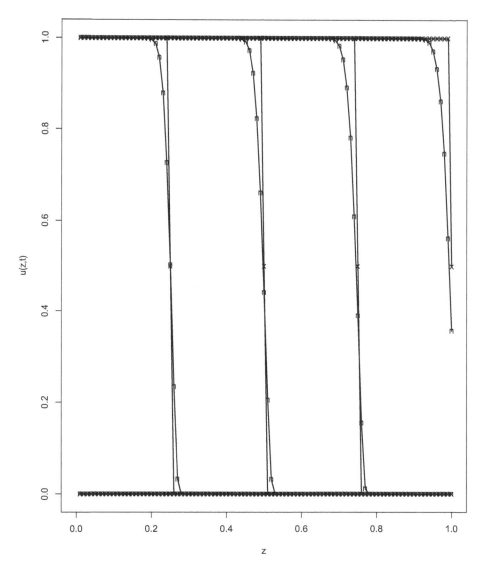

Figure A.2-1 $u(z,t)$ from eqs. (A.1), van Leer flux limiter

```
  smart=function(zl,zu,n,u,v) {
#
# Function smart computes the smart flux limiter
# approximation of a first derivative
#
# Declare arrays
```

```
  uz=rep(0,n);
 phi=rep(0,n);
   r=rep(0,n);
#
# Grid spacing
  dz=(zu-zl)/(n-1);
#
# Tolerance for limiter switching
  delta=1.0e-05;
#
# Positive v
  if(v >= 0){
    for(i in 3:(n-1)){
     if(abs(u[i]-u[i-1])<delta)
        phi[i]=0
      else{
        r[i]=(u[i+1]-u[i])/(u[i]-u[i-1])
        phi[i]=max(0,min(4,0.75*r[i]+0.25,2*r[i]))}
      if(abs(u[i-1]-u[i-2])<delta)
        phi[i-1]=0
      else{
        r[i-1]=(u[i]-u[i-1])/(u[i-1]-u[i-2])
        phi[i-1]=max(0,min(4,0.75*r[i-1]+0.25,2*r[i-1]))}
      flux2=u[i  ]+(u[i  ]-u[i-1])*phi[i  ]/2
      flux1=u[i-1]+(u[i-1]-u[i-2])*phi[i-1]/2
      uz[i]=(flux2-flux1)/dz
    }
    uz[1]=(-u[1]+u[2])/dz;
    uz[2]=(-u[1]+u[2])/dz;
    uz[n]=(u[n]-u[n-1])/dz;
  }
#
# Negative v
  if(v < 0){
    for(i in 2:(n-2)){
     if(abs(u[i]-u[i+1])<delta)
        phi[i]=0
      else{
        r[i]=(u[i-1]-u[i])/(u[i]-u[i+1])
        phi[i]=max(0,min(4,0.75*r[i]+0.25,2*r[i]))}
      if(abs(u[i+1]-u[i+2])<delta)
        phi[i+1]=0
      else{
        r[i+1]=(u[i]-u[i+1])/(u[i+1]-u[i+2])
```

Appendix A

```
          phi[i+1]=max(0,min(4,0.75*r[i+1]+0.25,2*r[i+1]))}
        flux2=u[i  ]+(u[i  ]-u[i+1])*phi[i  ]/2
        flux1=u[i+1]+(u[i+1]-u[i+2])*phi[i+1]/2
        uz[i]=-(flux2-flux1)/dz
    }
      uz[1]=(-u[1]+u[2])/dz;
      uz[n-1]=(-u[n-1]+u[n])/dz;
      uz[n  ]=(-u[n-1]+u[n])/dz;
  }
#
# All points concluded (z=zl,...,z=zu)
  return(c(uz));
}
```

Listing A.7: Function smart

smart is the same as vanl in Listing A.6 except for the limiting function phi listed next (for v>0).

```
#
# Positive v
  if(v >= 0){
    for(i in 3:(n-1)){
      if(abs(u[i]-u[i-1])<delta)
        phi[i]=0
      else{
        r[i]=(u[i+1]-u[i])/(u[i]-u[i-1])
        phi[i]=max(0,min(4,0.75*r[i]+0.25,2*r[i]))}
      if(abs(u[i-1]-u[i-2])<delta)
        phi[i-1]=0
      else{
        r[i-1]=(u[i]-u[i-1])/(u[i-1]-u[i-2])
        phi[i-1]=max(0,min(4,0.75*r[i-1]+0.25,2*r[i-1]))}
      flux2=u[i  ]+(u[i  ]-u[i-1])*phi[i  ]/2
      flux1=u[i-1]+(u[i-1]-u[i-2])*phi[i-1]/2
      uz[i]=(flux2-flux1)/dz
    }
      uz[1]=(-u[1]+u[2])/dz;
      uz[2]=(-u[1]+u[2])/dz;
      uz[n]=(u[n]-u[n-1])/dz;
  }
```

The code for v<0 is similar with the values u[i-1],u[i-2] replaced by u[i+1],u[i+2]. Mathematical and graphical details for the limiting function are available in [5], pp. 41–43.

The performance of the smart limiter is demonstrated by execution of the R routines in Listings A.1-5 with ncase=2. The numerical output is given in Table A.4.

[1] 5

[1] 102

ncase = 2

t	z	z-v*t	u(z,t)	uex(z,t)
0.00	0.00	0.00	1.000e+00	5.000e-01
0.00	0.10	0.10	0.000e+00	0.000e+00
0.00	0.20	0.20	0.000e+00	0.000e+00
0.00	0.30	0.30	0.000e+00	0.000e+00
0.00	0.40	0.40	0.000e+00	0.000e+00
0.00	0.50	0.50	0.000e+00	0.000e+00
0.00	0.60	0.60	0.000e+00	0.000e+00
0.00	0.70	0.70	0.000e+00	0.000e+00
0.00	0.80	0.80	0.000e+00	0.000e+00
0.00	0.90	0.90	0.000e+00	0.000e+00
0.00	1.00	1.00	0.000e+00	0.000e+00

t	z	z-v*t	u(z,t)	uex(z,t)
0.50	0.00	-0.50	1.000e+00	1.000e+00
0.50	0.10	-0.40	1.000e+00	1.000e+00
0.50	0.20	-0.30	1.000e+00	1.000e+00
0.50	0.30	-0.20	1.000e+00	1.000e+00
0.50	0.40	-0.10	9.999e-01	1.000e+00
0.50	0.50	0.00	4.544e-01	5.000e-01
0.50	0.60	0.10	-4.710e-07	0.000e+00
0.50	0.70	0.20	-1.156e-08	0.000e+00
0.50	0.80	0.30	-8.708e-11	0.000e+00
0.50	0.90	0.40	-3.044e-13	0.000e+00
0.50	1.00	0.50	-5.391e-16	0.000e+00

t	z	z-v*t	u(z,t)	uex(z,t)
1.00	0.00	-1.00	1.000e+00	1.000e+00
1.00	0.10	-0.90	1.000e+00	1.000e+00
1.00	0.20	-0.80	1.000e+00	1.000e+00
1.00	0.30	-0.70	1.000e+00	1.000e+00
1.00	0.40	-0.60	1.000e+00	1.000e+00
1.00	0.50	-0.50	1.000e+00	1.000e+00
1.00	0.60	-0.40	1.000e+00	1.000e+00
1.00	0.70	-0.30	1.000e+00	1.000e+00
1.00	0.80	-0.20	1.000e+00	1.000e+00
1.00	0.90	-0.10	9.999e-01	1.000e+00
1.00	1.00	0.00	4.600e-01	5.000e-01

Appendix A

```
ncall =   3352
```

Table A.4: Numerical output from Listings A.1-5, ncase=2, smart limiter

The following details about Table A.4 can be noted (with some repetition of the discussion of Table A.3).

- The dimensions of the solution matrix out returned by lsodes in the main program of Listing A.1 are again 5×102 as explained previously for Tables A.1, A.2, A.3
- The case is confirmed (as set in the main program of Listing A.1).

  ```
  ncase =   2
  ```

- The solution is displayed for z=0,1/100=0.01,...,1 as programmed in Listing A.1 (every tenth value of z is displayed as explained previously).
- IC (A.7-5) is confirmed ($t=0$). Note in particular the IC at $z - v_z t = 0$.
- The BC $u(z=0,t) = u_e = 1$ as programmed in Listing A.5 is confirmed.
- The computational effort as indicated by ncall = 3352 (less than for the van Leer limiter).

The graphical output is in Figure A.2-2. Comparison of the numerical solution $u(z,t)$ (plotted with n) and exact solution $h(z - v_z t)$ (plotted with x) indicates small numerical diffusion and no numerical oscillation, i.e., a major improvement over the FD solutions in Figures A.1-1, A.1-2.

Again, the exact solution $h(z - v_z t)$ is impossible to reproduce numerically since for $z - v_z t = 0$ the solution is undefined (infinite slope). With this in mind, the performance of the smart limiter is quite impressive.

This completes the discussion of the approximation of the derivative $\frac{\partial u}{\partial z}$ in eq. (A.1-1) by FDs and flux limiters. Discussion follows next for the approximation of the second derivative $\frac{\partial^2 u}{\partial z^2}$ in eqs. (A.4-1), (A.6-1).

A.4 SECOND-ORDER SPATIAL DERIVATIVES

From FD approximation (A.4-7),

$$\frac{\partial u}{\partial t} \approx D \left(\frac{\frac{\partial u}{\partial z}|_{z+\Delta z} - \frac{\partial u}{\partial z}|_z}{\Delta z} \right)$$

$$\approx D \left(\frac{\frac{u(z+\Delta z,t) - u(z,t)}{\Delta z} - \frac{u(z,t) - u(z-\Delta z,t)}{\Delta z}}{\Delta z} \right)$$

$$= D \left(\frac{u(z+\Delta z,t) - 2u(z,t) + u(z-\Delta z,t)}{\Delta z^2} \right) \quad \text{(A.9-1)}$$

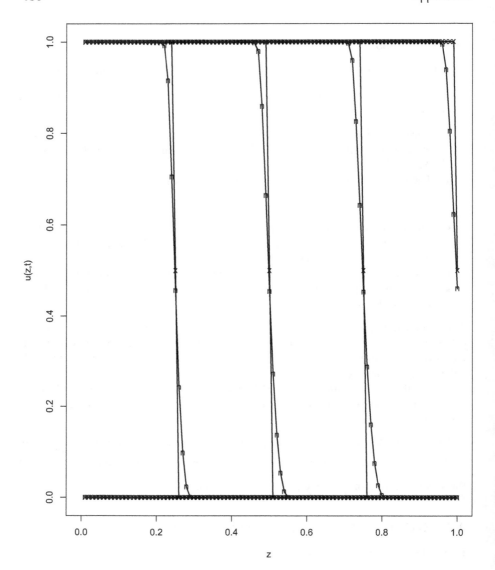

Figure A.2-2 $u(z,t)$ from eqs. (A.1), smart flux limiter

A FD approximation of the second derivative is therefore

$$\frac{\partial^2 u}{\partial z^2} \approx \frac{u(z+\Delta z,t) - 2u(z,t) + u(z-\Delta z,t)}{\Delta z^2}) + O(\Delta z^2) \qquad (A.9\text{-}2)$$

which is used in the following MOL analysis.

Appendix A

A main program for eqs. (A.4-1,2) and various BCs based on eqs. (A.9) follows.

```r
#
# Second order parabolic PDE
# (diffusion equation)
#
# Delete previous workspaces
  rm(list=ls(all=TRUE))
#
# Access ODE integrator
  library("deSolve");
#
# Access functions for numerical solution
  setwd("f:/Covid-19 neurological effects/appA");
  source("pde1c.R");
#
# Specify case
  ncase=1;
#
# Parameters
  nz=21;
  D=0.1;
  km=0.1;
 #
# Spatial grid in z
  zl=0;zu=1;dz=(zu-zl)/(nz-1);dzs=dz^2;
  z=seq(from=zl,to=zu,by=dz);
#
# Independent variable for ODE integration
  t0=0;tf=1;nout=11;
  tout=seq(from=t0,to=tf,by=(tf-t0)/(nout-1));
#
# Initial condition (t=0)
  u0=rep(0,nz);
  if(ncase==1){
    for(iz in 1:nz){
      u0[iz]=sin(pi/(zu-zl)*(z[iz]-zl));}
  }
  if(ncase==2){
    for(iz in 1:nz){
      u0[iz]=cos(pi/(zu-zl)*(z[iz]-zl));}
  }
  if(ncase==3){
    for(iz in 1:nz){
      u0[iz]=1;}
```

```
  }
  ncall=0;
#
# ODE integration
  out=lsodes(y=u0,times=tout,func=pde1c,
      sparsetype="sparseint",rtol=1e-6,
      atol=1e-6,maxord=5);
  nrow(out)
  ncol(out)
#
# Arrays for plotting numerical, exact solutions
    u=matrix(0,nrow=nz,ncol=nout);
  uex=matrix(0,nrow=nz,ncol=nout);
  if(ncase==1){
    for(it in 1:nout){
      for(iz in 1:nz){
         u[iz,it]=out[it,iz+1];
       uex[iz,it]=exp(-D*(pi/(zu-zl))^2*tout[it])*
                  sin(pi*(z[iz]-zl)/(zu-zl));
      }
      u[1,it]=0;
     u[nz,it]=0;
    }
  }
  if(ncase==2){
    for(it in 1:nout){
      for(iz in 1:nz){
         u[iz,it]=out[it,iz+1];
       uex[iz,it]=exp(-D*(pi/(zu-zl))^2*tout[it])*
                  cos(pi*(z[iz]-zl)/(zu-zl));
      }
    }
  }
  if(ncase==3){
    for(it in 1:nout){
      for(iz in 1:nz){
         u[iz,it]=out[it,iz+1];
      }
    }
  }
#
# Display numerical solution
# ncase=1,2
  if(ncase<3){
```

```
    iv=seq(from=1,to=nout,by=2);
     cat(sprintf("\n ncase = %2d\n",ncase));
    for(it in iv){
      cat(sprintf("\n     t      z       u(z,t)     uex(z,t)    error\n"));
      iv=seq(from=1,to=nz,by=10);
      for(iz in iv){
        err=u[iz,it]-uex[iz,it];
        cat(sprintf("%6.2f%6.2f%12.3e%12.3e%12.3e\n",
           tout[it],z[iz],u[iz,it],uex[iz,it],err));
      }
    }
  }
#
# ncase=3
  if(ncase==3){
    iv=seq(from=1,to=nout,by=2);
     cat(sprintf("\n ncase = %2d\n",ncase));
    for(it in iv){
      cat(sprintf("\n     t      z       u(z,t)\n"));
      iv=seq(from=1,to=nz,by=5);
      for(iz in iv){
        cat(sprintf("%6.2f%6.2f%12.3e\n",
           tout[it],z[iz],u[iz,it]));
      }
    }
  }
#
# Calls to ODE routine
  cat(sprintf("\n\n ncall = %5d\n\n",ncall));
#
# Plot PDE solution
#
# u
  par(mfrow=c(1,1));
  matplot(x=z,y=u,type="l",xlab="z",ylab="u(z,t)",
          xlim=c(zl,zu),lty=1,main="",lwd=2,col="black");
  matpoints(x=z,y=u,pch="n",lty=1,lwd=2,col="black");
#
# uex
  if(ncase<3){
  matpoints(x=z,y=uex,pch="x",lty=1,lwd=2,col="black");
  matpoints(x=z,y=uex,type="l",lty=1,lwd=2,col="black");
  }
```

Listing A.8: Main program for eqs. (A.4-1,2), parabolic

The following details about Listing A.8 can be noted (with some repetition of the discussion of Listing A.1).

- Previous workspaces are deleted.

  ```
  #
  # Second order parabolic PDE
  # (diffusion equation)
  #
  # Delete previous workspaces
    rm(list=ls(all=TRUE))
  ```

- The R ODE integrator library deSolve is accessed [6].

  ```
  #
  # Access ODE integrator
    library("deSolve");
  #
  # Access functions for numerical solution
    setwd("f:/Covid-19 neurological effects/appA");
    source("pde1c.R");
  ```

  Then the directory with the files for the solution of eqs. (A.4-1,2) is designated. Note that setwd (set working directory) uses / rather than the usual \.
  The ODE/MOL routine pde1c is discussed subsequently.

- An index for a case, with ncase=1,2,3, is specified. This index is used in the ODE/MOL routine pde1c considered subsequently.

  ```
  #
  # Specify case
    ncase=1;
  ```

- The parameters of eqs. (A.4-1,2).

  ```
  #
  # Parameters
    nz=21;
    D=0.1;
    km=0.1;
  ```

  where D, km are parameters in Robin BCs for ncase=3.

- A spatial grid for eq. (A.4-1) is defined with 21 points so that z = 0,1/20=0.05,...,1.

  ```
  #
  # Spatial grid in z
    zl=0;zu=1;dz=(zu-zl)/(nz-1);dzs=dz^2;
    z=seq(from=zl,to=zu,by=dz);
  ```

  The grid spacing is dz = Δz.

Appendix A

- An interval in t is defined with nout=11 output points.

```
#
# Independent variable for ODE integration
  t0=0;tf=1;nout=11;
  tout=seq(from=t0,to=tf,by=(tf-t0)/(nout-1));
```

- IC (A.4-2) is implemented for ncase=1,2,3.

```
# Initial condition (t=0)
  u0=rep(0,nz);
  if(ncase==1){
    for(iz in 1:nz){
      u0[iz]=sin(pi/(zu-zl)*(z[iz]-zl));}
  }
  if(ncase==2){
    for(iz in 1:nz){
      u0[iz]=cos(pi/(zu-zl)*(z[iz]-zl));}
  }
  if(ncase==3){
    for(iz in 1:nz){
      u0[iz]=1;}
  }
  ncall=0;
```

The three ICs are discussed subsequently. Also, the counter for the calls to pde1c is initialized.

- The system of nz=21 ODEs is integrated by the library integrator lsodes (available in deSolve, [6]). The inputs to lsodes are the ODE function, pde1c, the IC vector u0, and the vector of output values of t, tout. The length of u0 (nz = 21) informs lsodes how many ODEs are to be integrated. func,y,times are reserved names.

```
#
# ODE integration
  out=lsodes(y=u0,times=tout,func=pde1c,
      sparsetype="sparseint",rtol=1e-6,
      atol=1e-6,maxord=5);
  nrow(out)
  ncol(out)
```

nrow,ncol confirm the dimensions of out.

- $u(z,t)$ and the exact solution are placed in matrices for subsequent plotting.

```
#
# Arrays for plotting numerical, exact solutions
```

```r
    u=matrix(0,nrow=nz,ncol=nout);
  uex=matrix(0,nrow=nz,ncol=nout);
  if(ncase==1){
    for(it in 1:nout){
      for(iz in 1:nz){
         u[iz,it]=out[it,iz+1];
       uex[iz,it]=exp(-D*(pi/(zu-zl))^2*tout[it])*
                  sin(pi*(z[iz]-zl)/(zu-zl));
      }
      u[1,it]=0;
     u[nz,it]=0;
    }
  }
  if(ncase==2){
    for(it in 1:nout){
      for(iz in 1:nz){
         u[iz,it]=out[it,iz+1];
       uex[iz,it]=exp(-D*(pi/(zu-zl))^2*tout[it])*
                  cos(pi*(z[iz]-zl)/(zu-zl));
      }
    }
  }
  if(ncase==3){
    for(it in 1:nout){
      for(iz in 1:nz){
         u[iz,it]=out[it,iz+1];
      }
    }
  }
```

An exact (analytical) solution is included for ncase=1,2, and not for ncase=3, as discussed subsequently.
- The numerical and exact solutions are displayed for ncase=1,2.

```r
#
# Display numerical solution
# ncase=1,2
  if(ncase<3){
    iv=seq(from=1,to=nout,by=2);
     cat(sprintf("\n ncase = %2d\n",ncase));
    for(it in iv){
       cat(sprintf("\n    t      z      u(z,t)    uex(z,t)    error\n"));
       iv=seq(from=1,to=nz,by=5);
       for(iz in iv){
         err=u[iz,it]-uex[iz,it];
```

Appendix A

```
      cat(sprintf("%6.2f%6.2f%12.3e%12.3e%12.3e\n",
        tout[it],z[iz],u[iz,it],uex[iz,it],err));
      }
    }
  }
```

Every second value of t and every fifth value of z is selected with by=2,5.

- The numerical solution is displayed for ncase=3.

```
#
# ncase=3
  if(ncase==3){
    iv=seq(from=1,to=nout,by=2);
     cat(sprintf("\n ncase = %2d\n",ncase));
    for(it in iv){
      cat(sprintf("\n      t       z       u(z,t)\n"));
      iv=seq(from=1,to=nz,by=5);
      for(iz in iv){
        cat(sprintf("%6.2f%6.2f%12.3e\n",
          tout[it],z[iz],u[iz,it]));
      }
    }
  }
```

- The number of calls to pde1c at the end of the solutions is displayed.

```
#
# Calls to ODE routine
  cat(sprintf("\n\n ncall = %5d\n\n",ncall));
```

- The numerical solution is plotted with lines and points superimposed as n.

```
#
# Plot PDE solution
#
# u
  par(mfrow=c(1,1));
  matplot(x=z,y=u,type="l",xlab="z",ylab="u(z,t)",
          xlim=c(zl,zu),lty=1,main="",lwd=2,col="black");
  matpoints(x=z,y=u,pch="n",lty=1,lwd=2,col="black");
```

- The exact solutions for ncase=1,2 are plotted with lines and points superimposed as x.

```
#
# uex
```

```
      if(ncase<3){
      matpoints(x=z,y=uex,pch="x",lty=1,lwd=2,col="black");
      matpoints(x=z,y=uex,type="l",lty=1,lwd=2,col="black");
      }
```

This completes the discussion of the main program. The ODE/MOL routine pde1c is discussed next.

```
  pde1c=function(t,u,parm){
#
# Function pde1c computes the t derivative
# of u(z,t)
#
# BCs
  if(ncase==1){
     u[1]=0;
     u[nz]=0;}
#
# PDE
  ut=rep(0,nz);
  if(ncase==1){
    for(iz in 2:(nz-1)){
      ut[iz]=D*(u[iz+1]-2*u[iz]+u[iz-1])/dzs;
    }
     ut[1]=0;
     ut[nz]=0;
  }
  if(ncase==2){
    for(iz in 1:nz){
      if(iz==1){
        um=u[2];
        ut[1]=D*(u[2]-2*u[1]+um)/dzs;}
      if(iz==nz){
        up=u[nz-1];
        ut[nz]=D*(up-2*u[nz]+u[nz-1])/dzs;}
      if((iz>1)&(iz<nz)){
        ut[iz]=D*(u[iz+1]-2*u[iz]+u[iz-1])/dzs;}
    }
  }
  if(ncase==3){
    for(iz in 1:nz){
      if(iz==1){
         um=u[2]-(km/D)*(2*dz)*u[1];
         ut[1]=D*(u[2]-2*u[1]+um)/dzs;}
      if(iz==nz){
```

Appendix A

```
        up=u[nz-1]-(km/D)*(2*dz)*u[nz];
        ut[nz]=D*(up-2*u[nz]+u[nz-1])/dzs;}
     if((iz>1)&(iz<nz)){
        ut[iz]=D*(u[iz+1]-2*u[iz]+u[iz-1])/dzs;}
    }
  }
#
# Increment calls to pde1c
  ncall <<- ncall+1;
#
# Return derivative vector
  return(list(c(ut)));
  }
```

Listing A.9: ODE/MOL routine pde1c for eqs. (A.4-1,3,4)

The following details about Listing A.9 can be noted.

- The function is defined.

  ```
  pde1c=function(t,u,parm){
  #
  # Function pde1c computes the t derivative
  # of u(z,t)
  ```

  t is the current value of t in eq. (A.4-1). u is the 21-vector of ODE/PDE-dependent variables. parm is an argument to pass parameters to pde1c (unused, but required in the argument list). The arguments must be listed in the order stated to properly interface with lsodes called in the main program of Listing A.8. The derivative vector of the LHS of eq. (A.4-1) is calculated and returned to lsodes as explained subsequently.
- For ncase=1, homogeneous Dirichlet BCs are specified.

$$u(z=z_l,t) = u(z=z_u,t) = 0 \qquad (A.9\text{-}3)$$

Equation (A.4-1) is programmed for ncase=1.

```
#
# PDE
  ut=rep(0,nz);
  if(ncase==1){
    for(iz in 2:(nz-1)){
      ut[iz]=D*(u[iz+1]-2*u[iz]+u[iz-1])/dzs;
    }
     ut[1]=0;
    ut[nz]=0;
  }
```

The derivative $\dfrac{\partial^2 u}{\partial z^2}$ in eq. (A.4-1) is approximated with the three-point centered finite difference (FD) of eq. (A.9-2). For $z = z_l, z_u$, the BCs set the values of $u(z = z_l, t), u(z = z_u, t)$ and therefore the derivatives are set to zero to ensure the boundary values are maintained.

The MOL ODEs are defined at the interior grid poins in z,

```
if((iz>1)}&(iz<nz))
```

according to eq. (A.9-1).

- Equation (A.4-1) is programmed for ncase=2.

```
if(ncase==2){
  for(iz in 1:nz){
    if(iz==1){
      um=u[2];
      ut[1]=D*(u[2]-2*u[1]+um)/dzs;}
    if(iz==nz){
      up=u[nz-1];
      ut[nz]=D*(up-2*u[nz]+u[nz-1])/dzs;}
    if((iz>1)&(iz<nz)){
      ut[iz]=D*(u[iz+1]-2*u[iz]+u[iz-1])/dzs;}
  }
}
```

For a homogeneous Neumann BC at $z = z_l$,

$$\dfrac{\partial u(z = z_l, t)}{\partial z} = 0$$

the FD approximation is (iz=1)

$$\dfrac{\partial u(z_l, t)}{\partial z} \approx \dfrac{u(z_l + \Delta z, t) - u(z_l - \Delta z, t)}{2\Delta z} = 0$$

or $u(z_l - \Delta z, t) = u(z_l + \Delta z, t)$. $u(z_l - \Delta z, t)$ is an exterior (fictitious) value outside the grid in z (um in the preceding code).

Similarly, for a homogeneous Neumann BC at $z = z_u$,

$$\dfrac{\partial u(z = z_u, t)}{\partial z} = 0$$

the FD approximation is (iz=nz)

$$\dfrac{\partial u(z_u, t)}{\partial z} \approx \dfrac{u(z_u + \Delta z, t) - u(z_u - \Delta z, t)}{2\Delta z} = 0$$

or $u(z_u + \Delta z, t) = u(z_u - \Delta z, t)$. $u(z_u + \Delta z, t)$ is an exterior value outside the grid in z (up in the preceding code).

Appendix A

An extension of the homogeneous Neumann BCs can be considered as non-homogeneous (inhomogeneous, time varying) BCs. For example,

$$\frac{\partial u(z=z_l,t)}{\partial z} = g_{nl}(t) \quad \text{(A.9-4)}$$

$$\frac{\partial u(z=z_u,t)}{\partial z} = g_{nu}(t) \quad \text{(A.9-5)}$$

where $g_{nl}(t), g_{nu}(t)$ are functions to be specified. This case can be readily programmed since t is an input argument of the ODE/MOL routine (pde1c). The exterior values are

$$u(z=z_l-\Delta z,t) = u(z=z_l+\Delta z,t) - 2\Delta z g_{nl}(t) \quad \text{(A.9-6)}$$
$$u(z=z_u+\Delta z,t) = u(z=z_u-\Delta z,t) + 2\Delta z g_{nu}(t) \quad \text{(A.9-7)}$$

- Equation (A.4-1) is programmed for ncase=3.

```
if(ncase==3){
  for(iz in 1:nz){
    if(iz==1){
      um=u[2]-(km/D)*(2*dz)*u[1];
      ut[1]=D*(u[2]-2*u[1]+um)/dzs;}
    if(iz==nz){
      up=u[nz-1]-(km/D)*(2*dz)*u[nz];
      ut[nz]=D*(up-2*u[nz]+u[nz-1])/dzs;}
    if((iz>1)&(iz<nz)){
      ut[iz]=D*(u[iz+1]-2*u[iz]+u[iz-1])/dzs;}
  }
}
```

For a homogeneous Robin BC at $z=z_l$,

$$D\frac{\partial u(z=z_l,t)}{\partial z} - k_m u(z=z_l,t) = 0$$

the FD approximation is (iz=1)

$$D\frac{u(z_l+\Delta z,t) - u(z_l-\Delta z,t)}{2\Delta z} - k_m u(z_l,t) = 0$$

or $u(z_l-\Delta z,t) = u(z_l+\Delta z,t) - (k_m/D)(2\Delta z)u(z=z_l,t)$. $u(z_l-\Delta z,t)$ is an exterior (fictitious) value outside the grid in z (um in the preceding code). Similarly, for a homogeneous Robin BC at $z=z_u$,

$$D\frac{\partial u(z=z_u,t)}{\partial z} + k_m u(z=z_u,t) = 0$$

the FD approximation is (iz=nz)

$$D\frac{u(z_u+\Delta z,t) - u(z_u-\Delta z,t)}{2\Delta z} + k_m u(z_l,t) = 0$$

or $u(z_u+\Delta z,t) = u(z_u-\Delta z,t) - (k_m/D)(2\Delta z)u(z=z_u,t)$. $u(z_u+\Delta z,t)$ is an exterior value outside the grid in z (up in the preceding code).
The MOL ODEs are defined at the interior grid points in z,

```
if((iz>1)}&(iz<nz))
```

according to eq. (A.9-1).

An extension of the linear Robin BCs can be considered as nonlinear BCs. For example,

$$D\frac{\partial u(z=z_l,t)}{\partial z} - f_{rl}(u(z-z_l,t)) = 0$$

$$D\frac{\partial u(z=z_u,t)}{\partial z} + f_{ru}(u(z-z_u,t)) = 0$$

and f_{rl}, f_{ru} are functions to be specified. The FD approximations of the nonlinear Robin BCs are

$$D\frac{u(z_l+\Delta z,t) - u(z_l-\Delta z,t)}{2\Delta z} - f_{rl}(u(z-z_l,t)) = 0$$

$$D\frac{u(z_u+\Delta z,t) - u(z_u-\Delta z,t)}{2\Delta z} + f_{ru}(u(z-z_u,t)) = 0$$

or

$$u(z_l-\Delta z,t) = u(z_l+\Delta z,t) - (2\Delta z)f_{rl}(u(z=z_l,t)) \quad (A.9-8)$$

$$u(z_u+\Delta z,t) = u(z_u-\Delta z,t) - (2\Delta z)f_{ru}(u(z=z_u,t)) \quad (A.9-9)$$

Equations (A.9-8,9) can be programmed in the ODE/MOL routine (pde1c) since $u(z=z_l,t), u(z=z_u,t)$ are available in the input argument vector for $u(z,t)$.

- The counter for the calls to pde1c is incremented and returned to the main program of Listing A.8 by <<-.

```
#
# Increment calls to pde1c
  ncall <<- ncall+1;
```

- The vector ut is returned as a list as required by lsodes. c is the R vector utility. The final } concludes pde1c.

```
#
# Return derivative vector
  return(list(c(ut)));
  }
```

This completes the discussion of pde1c. The output from the main program of Listing A.8 and ODE/MOL routine pde1c of Listing A.9 is considered next, staring with ncase=1.

Appendix A

[1] 11

[1] 22

ncase = 1

t	z	u(z,t)	uex(z,t)	error
0.00	0.00	0.000e+00	0.000e+00	0.000e+00
0.00	0.25	7.071e-01	7.071e-01	0.000e+00
0.00	0.50	1.000e+00	1.000e+00	0.000e+00
0.00	0.75	7.071e-01	7.071e-01	0.000e+00
0.00	1.00	0.000e+00	1.225e-16	-1.225e-16

t	z	u(z,t)	uex(z,t)	error
0.20	0.00	0.000e+00	0.000e+00	0.000e+00
0.20	0.25	5.807e-01	5.804e-01	2.349e-04
0.20	0.50	8.212e-01	8.209e-01	3.323e-04
0.20	0.75	5.807e-01	5.804e-01	2.349e-04
0.20	1.00	0.000e+00	1.005e-16	-1.005e-16

t	z	u(z,t)	uex(z,t)	error
0.40	0.00	0.000e+00	0.000e+00	0.000e+00
0.40	0.25	4.769e-01	4.765e-01	3.873e-04
0.40	0.50	6.744e-01	6.738e-01	5.477e-04
0.40	0.75	4.769e-01	4.765e-01	3.873e-04
0.40	1.00	0.000e+00	8.252e-17	-8.252e-17

t	z	u(z,t)	uex(z,t)	error
0.60	0.00	0.000e+00	0.000e+00	0.000e+00
0.60	0.25	3.916e-01	3.911e-01	4.747e-04
0.60	0.50	5.538e-01	5.531e-01	6.714e-04
0.60	0.75	3.916e-01	3.911e-01	4.747e-04
0.60	1.00	0.000e+00	6.774e-17	-6.774e-17

t	z	u(z,t)	uex(z,t)	error
0.80	0.00	0.000e+00	0.000e+00	0.000e+00
0.80	0.25	3.216e-01	3.211e-01	5.196e-04
0.80	0.50	4.548e-01	4.540e-01	7.349e-04
0.80	0.75	3.216e-01	3.211e-01	5.196e-04
0.80	1.00	0.000e+00	5.560e-17	-5.560e-17

t	z	u(z,t)	uex(z,t)	error
1.00	0.00	0.000e+00	0.000e+00	0.000e+00
1.00	0.25	2.641e-01	2.635e-01	5.331e-04

```
1.00  0.50  3.735e-01  3.727e-01   7.540e-04
1.00  0.75  2.641e-01  2.635e-01   5.331e-04
1.00  1.00  0.000e+00  4.564e-17  -4.564e-17

ncall =    55
```

Table A.5: Numerical output from Listings A.8, A.9, ncase=1

The following details about Table A.5 can be noted (with some repetition of the discussion of Tables A.1-4).

- 11 t output points as the first dimension of the solution matrix out from lsodes as programmed in the main program of Listing A.8 (with nout=11).
- The solution matrix out returned by lsodes has 22 elements as a second dimension. The first element is the value of t. Elements 2 to 22 are $u(z,t)$ from eq. (A.4-1) (for each of the 21 output points).
- The solution is displayed for t=0,1/10=0.1,...,1 as programmed in Listing A.8 (every second value of t is displayed as explained previously).
- The solution is displayed for z=0,1/20=0.05,...,1 as programmed in Listing A.8 (every fifth value of z is displayed as explained previously).
- IC (A.4-10) is confirmed ($t = 0$).
- BCs (A.4-11,12) are confirmed ($z = z_l = 0, z = z_u = 1$).
- The computational effort as indicated by ncall = 55 is modest so that lsodes computed the solution to eq. (A.4-1) efficiently.

The graphical output is in Figures A.3-1. Comparison of the numerical solution $u(z,t)$ (plotted with n) and exact solution of eq. (A.4-18) (plotted with x) indicates good agreement between the numerical and exact solutions. In other words, Figure A.3-1 indicates the numerical integration of the parabolic (diffusion) PDE with Dirichlet BCs is straightforward in comparison with the first-order hyperbolic (advection) PDE.

The output for ncase=2 follows.

```
[1] 11

[1] 22

ncase =  2

   t     z     u(z,t)      uex(z,t)     error
0.00  0.00   1.000e+00    1.000e+00   0.000e+00
0.00  0.25   7.071e-01    7.071e-01   0.000e+00
0.00  0.50   6.123e-17    6.123e-17   0.000e+00
0.00  0.75  -7.071e-01   -7.071e-01   0.000e+00
```

```
 0.00  1.00  -1.000e+00   -1.000e+00    0.000e+00

   t     z     u(z,t)      uex(z,t)      error
 0.20  0.00   8.212e-01    8.209e-01    3.323e-04
 0.20  0.25   5.807e-01    5.804e-01    2.350e-04
 0.20  0.50  -2.565e-15    5.026e-17   -2.615e-15
 0.20  0.75  -5.807e-01   -5.804e-01   -2.350e-04
 0.20  1.00  -8.212e-01   -8.209e-01   -3.323e-04

   t     z     u(z,t)      uex(z,t)      error
 0.40  0.00   6.744e-01    6.738e-01    5.476e-04
 0.40  0.25   4.769e-01    4.765e-01    3.872e-04
 0.40  0.50  -7.886e-15    4.126e-17   -7.928e-15
 0.40  0.75  -4.769e-01   -4.765e-01   -3.872e-04
 0.40  1.00  -6.744e-01   -6.738e-01   -5.476e-04

   t     z     u(z,t)      uex(z,t)      error
 0.60  0.00   5.538e-01    5.531e-01    6.714e-04
 0.60  0.25   3.916e-01    3.911e-01    4.748e-04
 0.60  0.50   1.522e-14    3.387e-17    1.518e-14
 0.60  0.75  -3.916e-01   -3.911e-01   -4.748e-04
 0.60  1.00  -5.538e-01   -5.531e-01   -6.714e-04

   t     z     u(z,t)      uex(z,t)      error
 0.80  0.00   4.548e-01    4.540e-01    7.349e-04
 0.80  0.25   3.216e-01    3.211e-01    5.197e-04
 0.80  0.50   6.965e-14    2.780e-17    6.962e-14
 0.80  0.75  -3.216e-01   -3.211e-01   -5.197e-04
 0.80  1.00  -4.548e-01   -4.540e-01   -7.349e-04

   t     z     u(z,t)      uex(z,t)      error
 1.00  0.00   3.735e-01    3.727e-01    7.540e-04
 1.00  0.25   2.641e-01    2.635e-01    5.332e-04
 1.00  0.50   3.852e-14    2.282e-17    3.850e-14
 1.00  0.75  -2.641e-01   -2.635e-01   -5.332e-04
 1.00  1.00  -3.735e-01   -3.727e-01   -7.540e-04

ncall =    55
```

Table A.6: Numerical output from Listings A.8, A.9, ncase=2

The following details about Table A.6 can be noted (with some repetition of the discussion of Tables A.5).

- The dimensions of the array out from lsodes are again 11×22 (as for ncase=1, Table A.5).

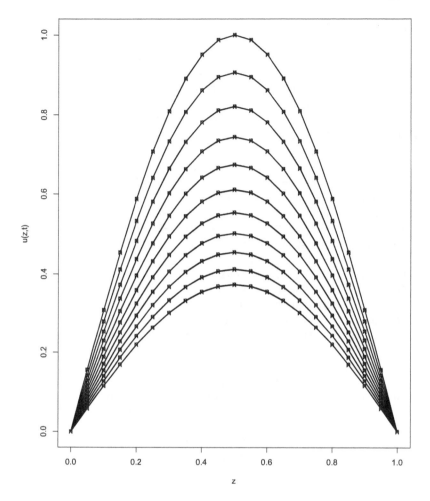

Figure A.3-1 $u(z,t)$ from eqs. (A.4), ncase=1, Dirichlet BCs

- The solution is displayed for t=0,1/10=0.1,...,1 as programmed in Listing A.8 (every second value of t) and z=0,1/20=0.05,...,1 (every fifth value of z).
- IC (A.4-13) is confirmed ($t = 0$).
- BCs (A.4-14,15) are confirmed ($z = z_l = 0, z = z_u = 1$).
- The computational effort as indicated by ncall = 55 is modest so that lsodes computed the solution to eq. (A.4-1) efficiently.

The graphical output is in Figures A.3-2. Comparison of the numerical solution $u(z,t)$ (plotted with n) and exact solution of eq. (A.4-19) (plotted with x) indicates good agreement between the numerical and exact solutions. In other words,

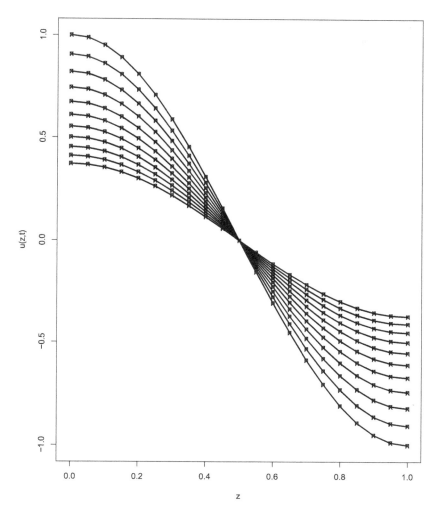

Figure A.3-2 $u(z,t)$ from eqs. (A.4), ncase=2, Neumann BCs

Figure A.3-2 indicates the numerical integration of the parabolic (diffusion) PDE with Neumann BCs is straightforward.

The output for ncase=3 follows.

[1] 11

[1] 22

ncase =   3

t	z	u(z,t)
0.00	0.00	1.000e+00
0.00	0.25	1.000e+00
0.00	0.50	1.000e+00
0.00	0.75	1.000e+00
0.00	1.00	1.000e+00

t	z	u(z,t)
0.20	0.00	3.355e-01
0.20	0.25	8.925e-01
0.20	0.50	9.892e-01
0.20	0.75	8.925e-01
0.20	1.00	3.355e-01

t	z	u(z,t)
0.40	0.00	2.547e-01
0.40	0.25	7.560e-01
0.40	0.50	9.148e-01
0.40	0.75	7.560e-01
0.40	1.00	2.547e-01

t	z	u(z,t)
0.60	0.00	2.122e-01
0.60	0.25	6.518e-01
0.60	0.50	8.123e-01
0.60	0.75	6.518e-01
0.60	1.00	2.122e-01

t	z	u(z,t)
0.80	0.00	1.822e-01
0.80	0.25	5.659e-01
0.80	0.50	7.118e-01
0.80	0.75	5.659e-01
0.80	1.00	1.822e-01

t	z	u(z,t)
1.00	0.00	1.580e-01
1.00	0.25	4.925e-01
1.00	0.50	6.212e-01
1.00	0.75	4.925e-01
1.00	1.00	1.580e-01

ncall =   145

Table A.7: Numerical output from Listings A.8, A.9, ncase=3

Appendix A

The following details about Table A.7 can be noted (with some repetition of the discussion of Table A.5).

- The dimensions of the array out from lsodes are again 11×22 (as for ncase=1,2, Tables A.5, A.6).
- The solution is displayed for t=0,1/10=0.1,...,1 as programmed in Listing A.8 (every second value of t) and z=0,1/20=0.05,...,1 (every fifth value of z).
- The IC $u(z,t=0) = 1$ (programmed in Listing A.8) is confirmed ($t = 0$) and in Figure A.3-3 that follows.
- BCs (A.4-16,17) are confirmed ($z = z_l = 0, z = z_u = 1$) and in Figure A.3-3 that follows.
- The computational effort as indicated by ncall = 145 is modest so that lsodes computed the solution to eq. (A.4-1) efficiently.

The graphical output is in Figures A.3-3. An exact solution to eq. (A.4-1), (A.4-16,17) is available, but is relatively complicated so it is not included in Figure A.3-3. The numerical solution of eqs. (A.4-1) (A.4-16,17) is straightforward, which indicates a principal feature of the numerical MOL.

This concludes the discussion of a parabolic PDE with eq. (A.4-1) (diffusion equation) as an example. Consideration is next for a multitype PDE, including convection and diffusion.

A.5 MULTITYPE PDE

Equation (A.6-1) has convection and diffusion terms ($v_z \neq 0, D \neq 0$) and is termed a convection-diffusion or hyperbolic-parabolic PDE. The MOL numerical integration (solution) of this multitype PDE is straightforward. A main program follows.

```
#
# Convection-diffusion PDE
#
# Delete previous workspaces
  rm(list=ls(all=TRUE))
#
# Access ODE integrator
  library("deSolve");
#
# Access functions for numerical solution
  setwd("f:/Covid-19 neurological effects/appA");
  source("pde1d.R");
  source("van1.R");
#
# Specify case
  ncase=1;
```

```
#
# Parameters
  vz=1;
  ue=1;
  if(ncase==1){
    D=1;
    nz=21;}
  if(ncase==2){
    D=0.001;
    nz=101;}
#
# Spatial grid in z
  zl=0;zu=1;dz=(zu-zl)/(nz-1);dzs=dz^2;
  z=seq(from=zl,to=zu,by=dz);
#
# Independent variable for ODE integration
  t0=0;tf=1;nout=11;
  tout=seq(from=t0,to=tf,by=(tf-t0)/(nout-1));
#
# Initial condition (t=0)
  u0=rep(0,nz);
  for(iz in 1:nz){
    u0[iz]=0}
  ncall=0;
#
# ODE integration
  out=lsodes(y=u0,times=tout,func=pde1d,
      sparsetype="sparseint",rtol=1e-6,
      atol=1e-6,maxord=5);
  nrow(out)
  ncol(out)
#
# Array for plotting numerical solution
  u=matrix(0,nrow=nz,ncol=nout);
  for(it in 1:nout){
    for(iz in 1:nz){
       u[iz,it]=out[it,iz+1];
    }
    u[1,it]=ue;
  }
#
# Display numerical solution
  Pe=vz*(zu-zl)/D;
  iv=seq(from=1,to=nout,by=5);
```

Appendix A

```
    cat(sprintf("\n  ncase = %2d    Pe = %7.3e\n",
                ncase,Pe));
    for(it in iv){
        cat(sprintf("\n     t        z       u(z,t)\n"));
        iv=seq(from=1,to=nz,by=10);
        for(iz in iv){
          cat(sprintf("%6.2f%6.2f%12.3e\n",
            tout[it],z[iz],u[iz,it]));}
    }
#
# Calls to ODE routine
    cat(sprintf("\n\n ncall = %5d\n\n",ncall));
#
# Plot PDE solution
#
# u
    par(mfrow=c(1,1));
    matplot(x=z[2:nz],y=u[2:nz,],type="l",xlab="z",ylab="u(z,t)",
            xlim=c(zl,zu),lty=1,main="",lwd=2,col="black");
    matpoints(x=z[2:nz],y=u[2:nz,],pch="n",lty=1,lwd=2,col="black");
```

Listing A.10: Main program for eqs. (A.6-1,2,3,5), hyperbolic-parabolic

The following details about Listing A.10 can be noted.

- Previous workspaces are deleted.

    ```
    #
    # Convection-diffusion PDE
    #
    # Delete previous workspaces
      rm(list=ls(all=TRUE))
    ```

- The R ODE integrator library deSolve is accessed [6].

    ```
    #
    # Access ODE integrator
      library("deSolve");
    #
    # Access functions for numerical solution
      setwd("f:/Covid-19 neurological effects/appA");
      source("pde1d.R");
      source("van1.R");
    ```

    Then the directory with the files for the solution of eqs. (A.6-1,2,3,5) is designated. Note that setwd (set working directory) uses / rather than the usual \.

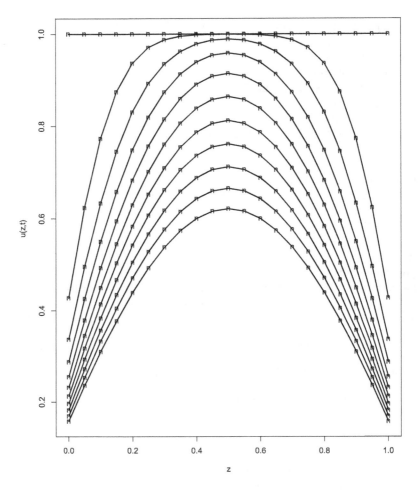

Figure A.3-3 $u(z,t)$ from eqs. (A.4), ncase=3, Robin BCs

The ODE/MOL routine pde1d is discussed subsequently. The van Leer flux limiter is used for the convection term, $v_z \dfrac{\partial u}{\partial z}$, in eq. (A.6-1).

- An index for a case, with ncase=1,2, is specified.

```
#
# Specify case
  ncase=1;
```

- The parameters of eqs. (A.6-1,2,3,5) are specified

```
#
# Parameters
```

Appendix A

```
    vz=1;
    ue=1;
    if(ncase==1){
      D=1;
      nz=21;}
    if(ncase==2){
      D=0.001;
      nz=101;}
```

The number of grid points in z, nz varies with ncase to provide adequate resolution of the graphical solution (in Figures A.4 discussed subsequently).

- A spatial grid for eq. (A.6-1) is defined with nz points.

```
#
# Spatial grid in z
  zl=0;zu=1;dz=(zu-zl)/(nz-1);dzs=dz^2;
  z=seq(from=zl,to=zu,by=dz);
```

The grid spacing is $dz = \Delta z$.

- The interval in t is defined with nout=11 output points.

```
#
# Independent variable for ODE integration
  t0=0;tf=1;nout=11;
  tout=seq(from=t0,to=tf,by=(tf-t0)/(nout-1));
```

- IC (A.6-2) is implemented, with $u(z, t = 0) = f(z) = 0$.

```
#
# Initial condition (t=0)
  u0=rep(0,nz);
  for(iz in 1:nz){
    u0[iz]=0;}
  ncall=0;
```

Also, the counter for the calls to pde1d is initialized.

- The system of nz ODEs is integrated by the library integrator lsodes (available in deSolve, [6]). As expected, the inputs to lsodes are the ODE function, pde1d, the IC vector u0, and the vector of output values of t, tout. The length of u0 (nz) informs lsodes how many ODEs are to be integrated. func,y,times are reserved names.

```
#
# ODE integration
```

```
  out=lsodes(y=u0,times=tout,func=pde1d,
      sparsetype="sparseint",rtol=1e-6,
      atol=1e-6,maxord=5);
  nrow(out)
  ncol(out)
```

nrow,ncol confirm the dimensions of out.

- The numerical values of $u(z,t)$ returned by lsodes are placed in matrix u. The BC $u(z = z_l,t)$=u[1,it]=ue is specified since it is not returned from lsodes (only solutions to ODEs are returned from lsodes).

```
#
# Array for plotting numerical solution
  u=matrix(0,nrow=nz,ncol=nout);
  for(it in 1:nout){
    for(iz in 1:nz){
       u[iz,it]=out[it,iz+1];
    }
    u[1,it]=ue;
  }
```

- The numerical values of $u(z,t)$ returned by lsodes are displayed. Every fifth value in t and every tenth value in z appear from by=5,10. The axial Peclet number, $P_e = \dfrac{v_z(z_u - z_l)}{D}$ is computed and displayed as explained subsequently.

```
#
# Display numerical solution
  Pe=vz*(zu-zl)/D;
  iv=seq(from=1,to=nout,by=5);
  cat(sprintf("\n ncase = %2d    Pe = %7.3e\n",
              ncase,Pe));
  for(it in iv){
      cat(sprintf("\n     t      z       u(z,t)\n"));
      iv=seq(from=1,to=nz,by=10);
      for(iz in iv){
         cat(sprintf("%6.2f%6.2f%12.3e\n",
           tout[it],z[iz],u[iz,it]));}
  }
```

- The number of calls to pde1d is displayed at the end of the solution.

```
#
# Calls to ODE routine
  cat(sprintf("\n\n ncall = %5d\n\n",ncall));
```

- The numerical solution is plotted as lines and superimposed points with the letter n.

Appendix A 161

```
#
# Plot PDE solution
#
# u
  par(mfrow=c(1,1));
  matplot(x=z[2:nz],y=u[2:nz,],type="l",xlab="z",
          ylab="u(z,t)",xlim=c(zl,zu),lty=1,main="",lwd=2,
          col="black");
  matpoints(x=z[2:nz],y=u[2:nz,],pch="n",lty=1,lwd=2,
            col="black");
```

The solution at $z = z_l$ is not plotted (`[2:nz,]`) to avoid the discontiniuty in the IC and BC, $u(z,t=0) = 0$, $u(z = z_l = 0, t) = u_e = 1$.

This completes the discussion of the main program of Listing A.10. The ODE/MOL routine, pde1d, called by lsodes in the main program is considered next.

```
  pde1d=function(t,u,parm){
#
# Function pde1d computes the t derivative
# of u(z,t)
#
# BC, z=zl
  u[1]=ue;
#
# PDE
  uz=vanl(zl,zu,nz,u,vz);
  ut=rep(0,nz);
  for(iz in 1:(nz-1)){
    if(iz==1){ut[iz]=0;}
    if(iz>1){
      ut[iz]=-vz*uz[iz]+
             D*(u[iz+1]-2*u[iz]+u[iz-1])/dzs;}
  }
#
# BC, z=zu
    ut[nz]=-vz*uz[nz];
#
# Increment calls to pde1d
  ncall <<- ncall+1;
#
# Return derivative vector
  return(list(c(ut)));
  }
```

Listing A.11: ODE/MOL routine pde1d for eqs. (A.6-1,2,3,5)

The following details about Listing A.11 can be noted.

- The function is defined.

    ```
    pde1d=function(t,u,parm){
    #
    # Function pde1d computes the t derivative
    # of u(z,t)
    ```

    t is the current value of t in eq. (A.6-1). u is the nz-vector of ODE/PDE-dependent variables. parm is an argument to pass parameters to pde1d (unused, but required in the argument list). The arguments must be listed in the order stated to properly interface with lsodes called in the main program of Listing A.10. The derivative vector of the LHS of eq. (A.6-1) is calculated and returned to lsodes as explained subsequently.
- BC (A.6-3) is programmed.

    ```
    #
    # BC, z=zl
      u[1]=ue;
    ```

- Equation (A.6-1) is programmed.

    ```
    #
    # PDE
      uz=van1(zl,zu,nz,u,vz);
      ut=rep(0,nz);
      for(iz in 1:(nz-1)){
        if(iz==1){ut[iz]=0;}
        if(iz>1){
          ut[iz]=-vz*uz[iz]+
                 D*(u[iz+1]-2*u[iz]+u[iz-1])/dzs;}
      }
    ```

    Note that the hyperbolic term -vz*uz[iz] and parabolic term D*(u[iz+1]-2*u[iz]+u[iz-1])/dzs are included.
- The dynamic BC (A.6-5) is programmed.

    ```
    #
    # BC, z=zu
      ut[nz]=-vz*uz[nz];
    ```

- The counter for the calls to pde1d is incremented and returned to the main program of Listing A.10 by <<-.

    ```
    #
    # Increment calls to pde1d
      ncall <<- ncall+1;
    ```

Appendix A

- The vector ut is returned as a list as required by lsodes. c is the R vector utility. The final } concludes pde1d.

```
#
# Return derivative vector
  return(list(c(ut)));
  }
```

The numerical and graphical output from the main program and ODE/MOL routine pde1d in Listings A.10, A.11 is considered next.

For ncase=1, the numerical output is in Table A.8.

[1] 11

[1] 22

```
ncase =  1    Pe = 1.000e+00

    t       z       u(z,t)
 0.00    0.00     1.000e+00
 0.00    0.50     0.000e+00
 0.00    1.00     0.000e+00

    t       z       u(z,t)
 0.50    0.00     1.000e+00
 0.50    0.50     7.066e-01
 0.50    1.00     3.360e-01

    t       z       u(z,t)
 1.00    0.00     1.000e+00
 1.00    0.50     8.376e-01
 1.00    1.00     6.300e-01

ncall =    247
```

Table A.8: Numerical output from Listings A.10, A.11, ncase=1

The following details about Table A.8 can be noted.

- The dimensions of the array out from lsodes are 11×22.
- The solution is displayed for t=0,1/10=0.1,...,1 as programmed in Listing A.10 (every fifth value of t) and z=0,1/20=0.05,...,1 (every tenth value of z).

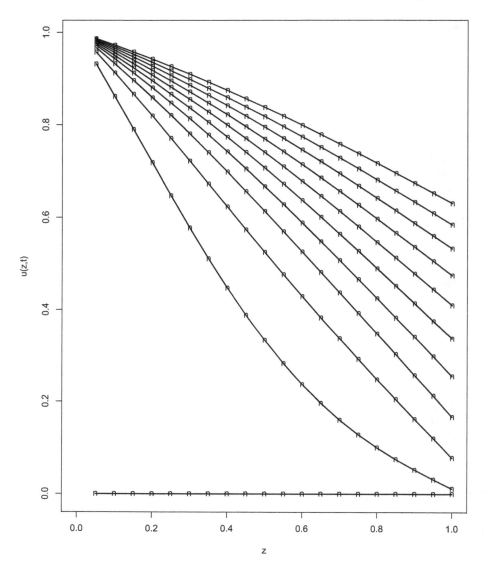

Figure A.4-1 $u(z,t)$ from eq. (A.6-1), ncase=1

- The IC $u(z, t = 0) = 0$ (programmed in Listing A.10) is confirmed ($t = 0$) and in Figure A.4-1 that follows.
- BC (A.6-3) is confirmed ($z = z_l = 0, z = z_u = 1$) and in Figure A.4-1 that follows.
- The computational effort as indicated by ncall = 247 is modest so that lsodes computed the solution to eq. (A.4-1) efficiently.

The graphical output is in Figures A.4-1. The solution for the strongly parabolic case, ncase=1, $P_e = \dfrac{v_z(z-z_l)}{D} = \dfrac{(1)(1)}{1} = 1$ (with $v_z = 1, D = 1$) is confirmed[5] (a smoothing of the discontinuity at $z = z_l = 0$).

For ncase=2, the numerical output is given in Table A.9.

[1] 11

[1] 102

```
ncase =  2    Pe = 1.000e+03

   t      z       u(z,t)
 0.00   0.00    1.000e+00
 0.00   0.10    0.000e+00
 0.00   0.20    0.000e+00
 0.00   0.30    0.000e+00
 0.00   0.40    0.000e+00
 0.00   0.50    0.000e+00
 0.00   0.60    0.000e+00
 0.00   0.70    0.000e+00
 0.00   0.80    0.000e+00
 0.00   0.90    0.000e+00
 0.00   1.00    0.000e+00

   t      z       u(z,t)
 0.50   0.00    1.000e+00
 0.50   0.10    1.000e+00
 0.50   0.20    1.000e+00
 0.50   0.30    1.000e+00
 0.50   0.40    9.969e-01
 0.50   0.50    5.362e-01
 0.50   0.60    6.452e-07
 0.50   0.70    3.982e-08
 0.50   0.80    9.385e-10
 0.50   0.90    7.827e-12
 0.50   1.00    2.390e-14

   t      z       u(z,t)
 1.00   0.00    1.000e+00
 1.00   0.10    1.000e+00
```

---

[5] The Peclet number, $P_e = \dfrac{v_z(z_u - z_l)}{D}$, can be considered as the ratio of the relative hyperbolic to parabolic contributions. For $P_e = 1$ (ncase=1), eq. (A.6-1) is strongly parabolic (diffusive). For $P_e = 1000$ (ncase=2), eq. (A.6-1) is strongly hyperbolic (convective).

```
1.00  0.20  1.000e+00
1.00  0.30  1.000e+00
1.00  0.40  1.000e+00
1.00  0.50  1.000e+00
1.00  0.60  1.000e+00
1.00  0.70  1.000e+00
1.00  0.80  9.999e-01
1.00  0.90  9.783e-01
1.00  1.00  5.299e-01

ncall =   5885
```

Table A.9: Numerical output from Listings A.10, A.11, ncase=2

The following details about Table A.9 can be noted.

- The dimensions of the array out from lsodes are 11×102.
- The solution is displayed for t=0,1/10=0.1,...,1 as programmed in Listing A.10 (every fifth value of t) and z=0,1/100=0.01,...,1 (every tenth value of z).
- The IC $u(z,t=0) = 0$ (programmed in Listing A.10) is confirmed ($t = 0$) and in Figure A.4-2 that follows.
- BC (A.6-3) is confirmed ($z = z_l = 0, z = z_u = 1$) and in Figure A.4-2 that follows.
- The computational effort indicated by ncall = 5885 reflects the computational requirement of the van Leer limiter.

The graphical output is in Figure A.4-2.

The solution for the strongly hyperbolic case, ncase=2, $P_e = \dfrac{v_z(z-z_l)}{D} = \dfrac{(1)(1)}{0.001} = 1000$ (with $v_z = 1, D = 0.001$) is confirmed, that is, propagation (movement left to right) of the discontinuity originating at $z = z_l = 0$. The solution has no numerical oscillation. The small axial diffusion from the van Leer flux limiter is physical (from eq. (A.6-1) with $D = 0.001$), not numerical. The right BC, eq. (A.6-5), gives a smooth exit of the moving front at $z = z_u = 1$.

This concludes the discussion of a hyperbolic-parabolic PDE with eq. (A.6-1) (convection-diffusion equation) as an example. Consideration is next for a 2×2 (two equations in two unknowns) system of simultaneous PDEs.

A.6 SIMULTANEOUS PDEs

The concluding PDE example is based on the application of eqs. (A.6-1,2,3,5), to a 2×2 PDE system. The equations are listed next.

$$\frac{\partial u_1}{\partial t} = -v_{z1}\frac{\partial u_1}{\partial z} + D_1\frac{\partial^2 u_1}{\partial z^2} + k_m(u_2 - u_1) \qquad \text{(A.10-1)}$$

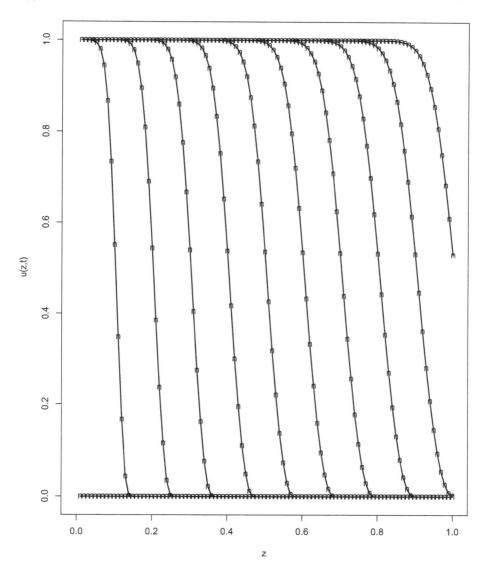

Figure A.4-2 $u(z,t)$ from eq. (A.6-1), ncase=2

$$\frac{\partial u_2}{\partial t} = -v_{z2}\frac{\partial u_2}{\partial z} + D_2\frac{\partial^2 u_2}{\partial z^2} - k_m(u_2 - u_1) \quad \text{(A.10-2)}$$

Equations (A.10-1,2) are first order in t and each requires an IC.

$$u_1(z,t=0) = f_1(z); \; u_2(z,t=0) = f_2(z) \quad \text{(A.10-3,4)}$$

where $f_1(z), f_2(z)$ are functions to be specified.

Equations (A.10-1,2) are second order in z and each requires two BCs. For a countercurrent system with $v_{z1} > 0, v_{z2} < 0$,

$$u_1(z=z_l,t) = g_1(t) \qquad (A.10\text{-}5)$$

$$\frac{\partial u_1(z=z_u,t)}{\partial t} + v_{z1}\frac{\partial u_1(z=z_u,t)}{\partial z} - k_m(u_2(z=z_u,t) - u_1(z=z_u,t)) = 0 \qquad (A.10\text{-}6)$$

$$u_2(z=z_u,t) = g_2(t) \qquad (A.10\text{-}7)$$

$$\frac{\partial u_2(z=z_l,t)}{\partial t} + v_{z2}\frac{\partial u_2(z=z_l,t)}{\partial z} + k_m(u_2(z=z_l,t) - u_1(z=z_l,t)) = 0 \qquad (A.10\text{-}8)$$

where $g_1(t), g_2(t)$ are functions to be specified.

A main program for eqs. (A.10) follows.

```
#
# Two convection-diffusion PDEs
#
# Delete previous workspaces
  rm(list=ls(all=TRUE))
#
# Access ODE integrator
  library("deSolve");
#
# Access functions for numerical solution
  setwd("f:/Covid-19 neurological effects/appA");
  source("pde1e.R");
  source("van1.R");
#
# Parameters
  vz1 =1;
  vz2=-1;
  u1e=1;
  u2e=0;
  D1=0.001;
  D2=0.001
  km=1;
  nz=41;
#
# Spatial grid in z
  zl=0;zu=1;dz=(zu-zl)/(nz-1);dzs=dz^2;
  z=seq(from=zl,to=zu,by=dz);
#
# Independent variable for ODE integration
  t0=0;tf=1;nout=11;
  tout=seq(from=t0,to=tf,by=(tf-t0)/(nout-1));
```

```
#
# Initial condition (t=0)
  u0=rep(0,2*nz);
  for(iz in 1:nz){
    u0[iz]=0;
    u0[iz+nz]=0;
}
  ncall=0;
#
# ODE integration
  out=lsodes(y=u0,times=tout,func=pde1e,
      sparsetype="sparseint",rtol=1e-6,
      atol=1e-6,maxord=5);
  nrow(out)
  ncol(out)
#
# Arrays for plotting numerical solution
  u1=matrix(0,nrow=nz,ncol=nout);
  u2=matrix(0,nrow=nz,ncol=nout);
  for(it in 1:nout){
    for(iz in 1:nz){
      u1[iz,it]=out[it,iz+1];
      u2[iz,it]=out[it,iz+1+nz];
    }
    u1[1,it] =u1e;
    u2[nz,it]=u2e;
  }
#
# Display numerical solution
  Pe1=abs(vz1)*(zu-zl)/D1;
  Pe2=abs(vz2)*(zu-zl)/D2;
  iv=seq(from=1,to=nout,by=5);
  cat(sprintf("\n Pe1 = %7.3e   Pe2 = %7.3e\n",Pe1,Pe2));
  for(it in iv){
      cat(sprintf("\n    t      z       u1(z,t)     u2(z,t)\n"));
      iv=seq(from=1,to=nz,by=10);
      for(iz in iv){
        cat(sprintf("%6.2f%6.2f%12.3e%12.3e\n",
          tout[it],z[iz],u1[iz,it],u2[iz,it]));}
  }
#
# Calls to ODE routine
  cat(sprintf("\n\n ncall = %5d\n\n",ncall));
#
```

```
# Plot PDE solution
#
# u1
  par(mfrow=c(1,1));
  matplot(x=z[2:nz],y=u1[2:nz,],type="l",xlab="z",
          ylab="u1(z,t)",xlim=c(zl,zu),lty=1,main="",lwd=2,
          col="black");
  matpoints(x=z[2:nz],y=u1[2:nz,],pch="n",lty=1,lwd=2,
            col="black");
#
# u2
  par(mfrow=c(1,1));
  matplot(x=z[1:(nz-1)],y=u2[1:(nz-1),],type="l",xlab="z",
          ylab="u2(z,t)",xlim=c(zl,zu),lty=1,main="",lwd=2,
          col="black");
  matpoints(x=z[1:(nz-1)],y=u2[1:(nz-1),],pch="n",lty=1,lwd=2,
            col="black");
```

Listing A.12: Main program for eqs. (A.10), 2×2 hyperbolic-parabolic

The following details about Listing A.10 can be noted.

- Previous workspaces are deleted.

  ```
  #
  # Two convection-diffusion PDEs
  #
  # Delete previous workspaces
    rm(list=ls(all=TRUE))
  ```

- The R ODE integrator library deSolve is accessed [6].

  ```
  #
  # Access ODE integrator
    library("deSolve");
  #
  # Access functions for numerical solution
    setwd("f:/Covid-19 neurological effects/appA");
    source("pde1e.R");
    source("vanl.R");
  ```

  Then the directory with the files for the solution of eqs. (A.10) is designated. Note that setwd (set working directory) uses / rather than the usual \.
  The ODE/MOL routine pde1e is discussed subsequently. The van Leer flux limiter is used for the convection terms, $v_{z1}\dfrac{\partial u_1}{\partial z}$, $v_{z2}\dfrac{\partial u_2}{\partial z}$ in eqs. (A.10-1,2).

Appendix A

- The parameters of eqs. (A.10) are specified

```
#
# Parameters
  vz1 =1;
  vz2=-1;
  u1e=1;
  u2e=0;
  D1=0.001;
  D2=0.001
  km=1;
  nz=41;
```

The following details about these parameters can be noted.
- Countercurrent convection is specified with velocities of opposite sign, vz1=1, vz2=-1.
- The boundary value $u_1(z=z_l,t)=u_{e1}=1$ moves the PDE system away from the ICs.
- The PDE is strongly convective with D1=D2=0.001 so that $P_{e1}=P_{e2}=1000$.
- The transfer coefficient in eqs. (A.10-1,2) is specified, $k_m = 1$.
- The number of grid points in z, nz=41, provides adequate spatial resolution of the graphical solution (in Figures A.5 discussed subsequently).

- A spatial grid for eqs. (A.10-1,2) is defined with nz=41 points.

```
#
# Spatial grid in z
  zl=0;zu=1;dz=(zu-zl)/(nz-1);dzs=dz^2;
  z=seq(from=zl,to=zu,by=dz);
```

The grid spacing is dz = Δz.
- The interval in t is defined with *nout* = 11 output points.

```
#
# Independent variable for ODE integration
  t0=0;tf=1;nout=11;
  tout=seq(from=t0,to=tf,by=(tf-t0)/(nout-1));
```

- ICs (A.10-3,4) are implemented, with $u_1(z,t=0) = f_1(z) = 0$. $u_2(z,t=0) = f_2(z) = 0$.

```
#
# Initial condition (t=0)
  u0=rep(0,2*nz);
  for(iz in 1:nz){
```

```
      u0[iz]=0;
      u0[iz+nz]=0;
    }
    ncall=0;
```

Also, the counter for the calls to pde1e is initialized.

- The system of 2*nz=82 ODEs is integrated by the library integrator lsodes (available in deSolve, [6]). As expected, the inputs to lsodes are the ODE function, pde1e, the IC vector u0, and the vector of output values of t, tout. The length of u0 (2*nz=82) informs lsodes how many ODEs are to be integrated. func,y,times are reserved names.

```
# ODE integration
  out=lsodes(y=u0,times=tout,func=pde1e,
      sparsetype="sparseint",rtol=1e-6,
      atol=1e-6,maxord=5);
  nrow(out)
  ncol(out)
```

nrow,ncol confirm the dimensions of out.

- The numerical values of $u_1(z,t), u_2(z,t)$ returned by lsodes are placed in matrix u1,u2. The BCs $u_1(z = z_l, t)$=u1[1,it]=u1e, $u_1(z = z_u, t)$=u2[nz,it]=u2e are specified since they are not returned from lsodes (only solutions to ODEs are returned from lsodes).

```
#
# Arrays for plotting numerical solution
  u1=matrix(0,nrow=nz,ncol=nout);
  u2=matrix(0,nrow=nz,ncol=nout);
  for(it in 1:nout){
    for(iz in 1:nz){
      u1[iz,it]=out[it,iz+1];
      u2[iz,it]=out[it,iz+1+nz];
    }
    u1[1,it] =u1e;
    u2[nz,it]=u2e;
  }
```

- The numerical values of $u_1(z,t), u_2(z,t)$ returned by lsodes are displayed. Every fifth value in t and every tenth value in z appear from by=5,10. The axial Peclet numbers, $P_{e1} = P_{e2} = 1000$ are computed and displayed as explained subsequently.

```
#
# Display numerical solution
  Pe1=abs(vz1)*(zu-zl)/D1;
```

```
    Pe2=abs(vz2)*(zu-zl)/D2;
    iv=seq(from=1,to=nout,by=5);
    cat(sprintf("\n Pe1 = %7.3e    Pe2 = %7.3e\n",Pe1,Pe2));
    for(it in iv){
        cat(sprintf("\n   t     z     u1(z,t)    u2(z,t)\n"));
        iv=seq(from=1,to=nz,by=10);
        for(iz in iv){
            cat(sprintf("%6.2f%6.2f%12.3e%12.3e\n",
                tout[it],z[iz],u1[iz,it],u2[iz,it]));}
    }
```

- The number of calls to pde1e is displayed at the end of the solution.

```
#
# Calls to ODE routine
    cat(sprintf("\n\n ncall = %5d\n\n",ncall));
```

- The numerical solution $u_1(z,t)$ is plotted as lines and superimposed points with the letter n.

```
#
# Plot PDE solution
#
# u1
    par(mfrow=c(1,1));
    matplot(x=z[2:nz],y=u1[2:nz,],type="l",xlab="z",
            ylab="u1(z,t)",xlim=c(zl,zu),lty=1,main="",lwd=2,
            col="black");
    matpoints(x=z[2:nz],y=u1[2:nz,],pch="n",lty=1,lwd=2,
              col="black");
```

The solution at $z = z_l$ is not plotted ([2:nz,]) to avoid the discontiniuty in the IC and BC, $u_1(z,t=0) = 0$, $u_1(z = z_l = 0, t) = u_{e1} = 1$.

- The numerical solution $u_2(z,t)$ is plotted as lines and superimposed points with the letter n.

```
#
# u2
    par(mfrow=c(1,1));
    matplot(x=z[1:(nz-1)],y=u2[1:(nz-1),],type="l",xlab="z",
            ylab="u2(z,t)",xlim=c(zl,zu),lty=1,main="",lwd=2,
            col="black");
    matpoints(x=z[1:(nz-1)],y=u2[1:(nz-1),],pch="n",lty=1,
              lwd=2,col="black");
```

The solution at $z = z_u$ is not plotted ([1:(nz-1)]) to avoid a possible discontinuity in the IC and BC, $u_2(z,t=0)$, $u_2(z = z_u, t)$.

This completes the discussion of the main program of Listing A.12. The ODE/MOL routine, pde1e, called by lsodes in the main program, is considered next.

```
  pde1e=function(t,u,parm){
#
# Function pde1e computes the t derivative
# of u1(z,t), u2(z,t)
#
# One vector to two vectors
  u1=rep(0,nz);
  u2=rep(0,nz);
  for(iz in 1:nz){
    u1[iz]=u[iz];
    u2[iz]=u[iz+nz];
  }
#
# BCs, z=zl, z=zu
  u1[1] =u1e;
  u2[nz]=u2e;
#
# PDEs
#
# u1t
  u1z=vanl(zl,zu,nz,u1,vz1);
  u1t=rep(0,nz);
  for(iz in 1:(nz-1)){
    if(iz==1){u1t[1]=0;}
    if(iz>1){
      u1t[iz]=-vz1*u1z[iz]+
              D1*(u1[iz+1]-2*u1[iz]+u1[iz-1])/dzs+
              km*(u2[iz]-u1[iz]);}
  }
#
# u2t
  u2z=vanl(zl,zu,nz,u2,vz2);
  u2t=rep(0,nz);
  for(iz in 2:nz){
    if(iz==nz){u2t[nz]=0;}
    if(iz<nz){
      u2t[iz]=-vz2*u2z[iz]+
              D2*(u2[iz+1]-2*u2[iz]+u2[iz-1])/dzs-
              km*(u2[iz]-u1[iz]);}
  }
#
```

```
# BC, z=zl, zu
   u1t[nz]=-vz1*u1z[nz]+km*(u2[nz]-u1[nz]);
   u2t[1] =-vz2*u2z[1] -km*(u2[1]-u1[1]);
#
# Two vectors to one vector
   ut=rep(0,2*nz);
   for(iz in 1:nz){
     ut[iz]    =u1t[iz];
     ut[iz+nz]=u2t[iz];
   }
#
# Increment calls to pde1e
   ncall <<- ncall+1;
#
# Return derivative vector
   return(list(c(ut)));
   }
```

Listing A.13: ODE/MOL routine pde1e for eqs. (A.10), 2×2 hyperbolic-parabolic

The following details about Listing A.11 can be noted.

- The function is defined.

  ```
  pde1e=function(t,u,parm){
  #
  # Function pde1e computes the t derivative
  # of u1(z,t), u2(z,t)
  ```

  t is the current value of t in eqs. (A.10-1,2). u is the 2*nz=82-vector of ODE/PDE dependent variables. parm is an argument to pass parameters to pde1e (unused, but required in the argument list). The arguments must be listed in the order stated to properly interface with lsodes called in the main program of Listing A.12. The derivative vector of the LHS of eqs. (A.10-1,2) is calculated and returned to lsodes as explained subsequently.
- u is placed in two vectors, u1, u2, to facilitate the programming of eqs. (A.10-1,2)

  ```
  #
  # One vector to two vectors
    u1=rep(0,nz);
    u2=rep(0,nz);
    for(iz in 1:nz){
      u1[iz]=u[iz];
      u2[iz]=u[iz+nz];
    }
  ```

- BCs (A.10-5,7) are programmed.

  ```
  #
  # BCs, z=zl, z=zu
    u1[1] =u1e;
    u2[nz]=u2e;
  ```

  iz=1,nz specify points $z = z_l, z_u$, respectively (programmed in Listing A.12). u1e=1 moves the PDE solutions away from tne homogeneous ICs, eqs. (A.10-3,4) with $f_1(z) = f_2(z) = 0$ programmed in Listing A.12.
- Equation (A.10-1) is programmed.

  ```
  #
  # PDEs
  #
  # u1t
    u1z=vanl(zl,zu,nz,u1,vz1);
    u1t=rep(0,nz);
    for(iz in 1:(nz-1)){
      if(iz==1){u1t[1]=0;}
      if(iz>1){
        u1t[iz]=-vz1*u1z[iz]+
              D1*(u1[iz+1]-2*u1[iz]+u1[iz-1])/dzs+
              km*(u2[iz]-u1[iz]);}
    }
  ```

  Note that the hyperbolic term -vz1*u1z[iz] and parabolic term D1*(u1[iz+1]-2*u1[iz]+u1[iz-1])/dzs are included.
- Equation (A.10-2) is programmed.

  ```
  #
  # u2t
    u2z=vanl(zl,zu,nz,u2,vz2);
    u2t=rep(0,nz);
    for(iz in 2:nz){
      if(iz==nz){u2t[nz]=0;}
      if(iz<nz){
        u2t[iz]=-vz2*u2z[iz]+
              D2*(u2[iz+1]-2*u2[iz]+u2[iz-1])/dzs-
              km*(u2[iz]-u1[iz]);}
    }
  ```

  The hyperbolic term -vz2*u2z[iz] and parabolic term D2*(u2[iz+1]-2*u2[iz]+u2[iz-1])/dzs are included.
- The dynamic BCs (A.10-6,8) are programmed. iz=1,nz specify points $z = z_l, z_u$, respectively (programmed in Listing A.12).

Appendix A   177

```
#
# BC, z=zl, zu
  u1t[nz]=-vz1*u1z[nz]+km*(u2[nz]-u1[nz]);
  u2t[1]  =-vz2*u2z[1]  -km*(u2[1]-u1[1]);
```

- The counter for the calls to pde1e is incremented and returned to the main program of Listing A.13 by <<-.

```
#
# Increment calls to pde1e
  ncall <<- ncall+1;
```

- The vector ut is returned as a list as required by lsodes. c is the R vector utility. The final } concludes pde1e.
- The derivative vectors u1t, u2t are placed in a single vector ut to return to lsodes for the next step along the solution.

```
#
# Two vectors to one vector
  ut=rep(0,2*nz);
  for(iz in 1:nz){
    ut[iz]    =u1t[iz];
    ut[iz+nz]=u2t[iz];
  }
```

- The vector ut is returned as a list as required by lsodes. c is the R vector utility. The final } concludes pde1e.

```
#
# Return derivative vector
  return(list(c(ut)));
  }
```

The numerical and graphical output from the main program and ODE/MOL routine pde1e in Listings A.12, A.13 is considered next. The numerical output is in Table A.10.

[1] 11

[1] 83

Pe1 = 1.000e+03   Pe2 = 1.000e+03

```
    t      z     u1(z,t)      u2(z,t)
 0.00   0.00   1.000e+00    0.000e+00
 0.00   0.25   0.000e+00    0.000e+00
```

```
 0.00    0.50    0.000e+00    0.000e+00
 0.00    0.75    0.000e+00    0.000e+00
 0.00    1.00    0.000e+00    0.000e+00

   t       z      u1(z,t)      u2(z,t)
 0.50    0.00    1.000e+00    1.994e-01
 0.50    0.25    8.007e-01    8.846e-02
 0.50    0.50    3.436e-01    6.350e-03
 0.50    0.75    2.056e-08    5.081e-10
 0.50    1.00    7.773e-11    0.000e+00

   t       z      u1(z,t)      u2(z,t)
 1.00    0.00    1.000e+00    3.271e-01
 1.00    0.25    8.324e-01    2.189e-01
 1.00    0.50    6.688e-01    1.282e-01
 1.00    0.75    5.131e-01    5.549e-02
 1.00    1.00    2.119e-01    0.000e+00

ncall =  3969
```

Table A.10: Numerical output from Listings A.12, A.13

The following details about Table A.10 can be noted.

- The dimensions of the array out from lsodes are 11×83 corresponding to nout=11, 2(nz)+1=2(41)+1=83.
- The solution is displayed for t=0,1/10=0.1,...,1 as programmed in Listing A.12 (every fifth value of t) and z=0,1/40=0.025,...,1 (every tenth value of z).
- The ICs $u(z, t = 0) = u_2(z, t = 0) = 0$ (programmed in Listing A.12) are confirmed ($t = 0$) and in Figures A.5-1,2 that follow.
- BCs (A.10-5,7) are confirmed, $(u_1(z = z_l = 0, t) = 1, u_2(z = z_u = 1, t) = 0$, and in Figures A.5-1,2 that follow. The numerical solutions $u_1(z,t), u_2(z,t)$ are the response to BC (A.10-5) starting from homogeneous ICs (A.10-3,4).
- The computational effort indicated by ncall = 3969 reflects the computational requirement of the van Leer limiter.

The graphical output is in Figures A.5-1,2.

$u_1(z,t)$ is the response to BC (A.10-5) starting from homogeneous IC (A.10-3), and includes the effect of the transfer term $k_m(u_2(z,t) - u_1(z,t))$. $u_2(z,t)$ is the response to the transfer term $k_m(u_2(z,t) - u_1(z,t))$ starting from homogeneous IC (A.10-4),

Appendix A

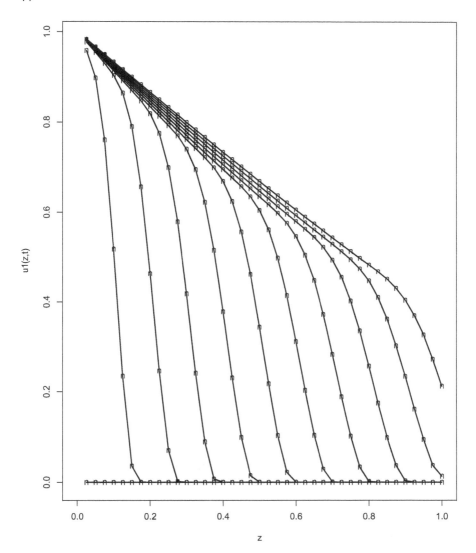

Figure A.5-1 $u_1(z,t)$ from eq. (A.10-1)

A.7 SUMMARY AND CONCLUSIONS

This appendix is intended as a basic introduction to PDE analysis, with emphasis on evolutionary (time dependent) PDEs. PDE terminology and geometric classification are presented, followed by computer implementation of PDE systems by the method

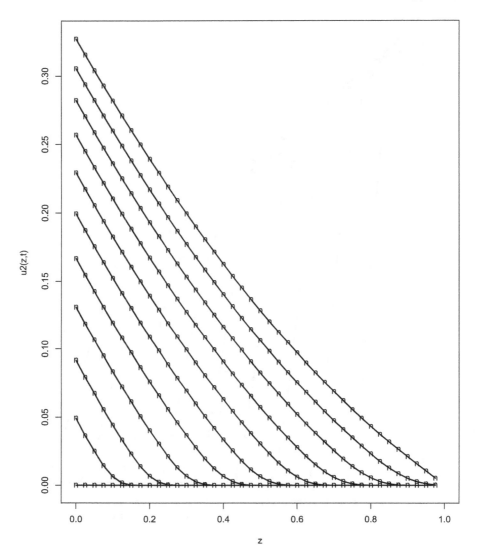

Figure A.5-2 $u_2(z,t)$ from eq. (A.10-2)

of lines. Source routines in R with detailed documentation provide a starting point for the development of new PDE models.

REFERENCES

1. https://en.wikipedia.org/wiki/Cartesian‿coordinate‿system.
2. https://en.wikipedia.org/wiki/Cylindrical‿coordinate‿system.

3. https://en.wikipedia.org/wiki/Spherical`coordinate`system.
4. https://en.wikipedia.org/wiki/Curvilinear`coordinates.
5. Griffiths, G.W., and W.E. Schiesser (2012), *Traveling Wave Analysis of Partial Differential Equations*, Academic Press-Elsevier, Oxford, UK.
6. Soetaert, K., J. Cash, and F. Mazzia (2012), *Solving Differential Equations in R*, Springer-Verlag, Heidelberg, Germany.

Index

advection 1–3, 5, 7–8, 10–11, 14, 17, 21, 24, 29, 40, 43, 46, 79
advection PDE 100–101

BBB *see* blood brain barrier
blood brain barrier 1–2, 8, 15, 17, 38
blood convection 1–3, 5, 7–8, 10, 14
blood oxygen concentration 1–3, 10, 15–16, 46, 55
    time variation 55
blood to brain tissue transfer 17–19, 29
boundary condition (BC) 2, 9–12, 14, 17, 28–29, 31, 46, 55, 57, 79, 100–101
    Dirichlet 101
    Neumann 101
    Robin 101
brain cognitive impairment 1, 38, 67
brain fog 1

capillary blood flow 1–3, 5, 7–8, 10–11, 14–16, 40, 43, 46, 48
Cartesian coordinates 17, 100
cocurrent PDEs 168
cognitive impairment 1, 38, 67
convection 1–3, 5, 7–8, 10–11, 14–16, 40, 43, 46, 48, 55, 57
    divergence operator 105
convection diffusion reaction equation 107
countercurrent PDEs 168
Covid-19 1
    impaired respiratory function 1, 58
cylindrical coordinates 100

diffusion 17–19, 21, 24, 28, 42, 47, 55
diffusion equation 102, 144

analytical solution 105–106, 142, 149–153
    ODE/MOL routine 144–149
diffusivity 17–19, 21, 24, 28, 42, 47, 55, 57, 79
Dirichlet boundary condition 101–102, 105, 149–152
discontinuity 14, 23, 27, 39, 41, 51, 55–56, 80
divergence operator 103–104
    Cartesian coordinates 103–104
    convection 105
    cylindrical coordinates 103–104
    spherical coordinates 103–104

elliptic PDE 106

false transients 107
fatique 1
Fick's first law 18, 103
Fick's second law 102
finite difference (FD) 3, 12, 28–29
    noncentered 115, 117
    order 12, 30
    three point centered 30–31, 47, 49, 77, 79, 122, 136, 146–148
    two point centered 30–31, 47, 49, 77, 79, 116–117, 146–148
    two point upwind 29–31, 46, 48, 77, 79, 108, 116
flux limiters 123
    smart 123–125, 131–136
    van Leer 123–131

geometric PDE classification 100
    elliptic 106
    hyperbolic 100

geometric PDE classification (*cont.*)
  hyperbolic-parabolic 107, 165
  parabolic 102

Heaviside function 101, 108, 110–111, 113
hyperbolic PDE 100
  first order 100–101
    boundary condition (BC) 100
    initial condition (IC) 100
    main program 109
    ODE/MOL routine 109, 115
  second order 101
    boundary conditions 101
    expressed as first order 101–102
    initial conditions 101
hyperbolic-parabolic PDE 107, 165
hypoxia 1, 7

impaired respiratory function 1, 58
initial condition (IC) 2, 8, 17, 39, 100–101

Lagrangian coordinate 100–101, 113
Laplace's equation 106
Laplacian operator 104–105
  Cartesian coordinates 104
  cylindrical operator 105
  spherical coordinates 105
long Covid 1
LHS PDE derivatives 68
  main program 77
lungs 1
  impaired respiratory function 1, 58

main program 5
mass transfer coefficient 1–3, 5, 7–8, 10–11, 14–15, 24, 40, 55, 57
mass transfer rate 1–3, 10
method of lines *see* MOL
mixed derivative 99, 102
MOL 5–6, 10, 15, 38, 108
  ODE/MOL routine 5–6, 8, 10–11, 22, 27

time derivative vector 1–3, 11, 28, 32
Neumann boundary condition 17, 101, 106, 153
neurological effects 1
  brain cognitive impairment 1, 38, 67
  brain fog 1
  fatique 1
neuron 1
  density 1, 47
  time derivative 39, 47, 49
  rate constant 39–40, 43, 47, 49, 53, 78
one PDE model 1
  blood oxygen concentration 1, 10, 15–16
  boundary condition (BC) 2, 9–12, 14
  convection 1–3, 5, 7–8, 10–11, 14
  dependent variable
  blood oxygen concentration 1–3, 10, 15–16
  discontinuity 14
  finite difference (FD) 12
  order 12
  graphical output 15–16
  independent variables
  space 8
  time 8
  initial condition (IC) 2, 8
  main program 5
  mass transfer coefficient 1–3, 5, 7, 10–11, 14–15, 18–19
  numerical output 13
  ODE/MOL routine 5–6, 8, 10–11
  spatial derivative 1–3
  spatial grid 8
  spatiotemporal modeling 1–3
  time derivative 1–3, 11
  time interval 8
  velocity 1–3, 5, 7–8, 10–11, 14

Index

ODE/MOL routine 5–6, 8, 10–11
oxygen concentration 1, 15–16
   along capillary 1, 15–16

parabolic PDE 102, 137–144
   ODE/MOL routine 144–149
partial differential equation *see* PDE
PDE *see also* one, two, three PDE
   advecion equation 100–101
   Cartesian coordinates 17, 100
   convection diffusion reaction 107
   cylindrical coordinates 100
   diffusion equation 102, 144
   analytical solution 105–106, 142, 149–153
   ODE/MOL routine 144–149
   Dirichlet boundary condition 101–102, 105, 149–152
   divergence operator 103–104
   Cartesian coordinates 103–104
   cylindrical coordinates 103–104
   spherical coordinates 103–104
   convection 105
   elliptic PDE 106
   false transients 107
   Fick's first law 18, 103
   Fick's second law 102
   finite difference (FD) 3, 12, 28–29
   noncentered 115, 117
   order 12, 30
   three point centered 30–31, 47, 49, 77, 79, 122, 136, 146–148
   two point centered 30–31, 47, 49, 77, 79, 116–117, 146–148
   two point upwind 29–31, 46, 48, 77, 79, 108, 121
   first order hyperbolic 100–101
   boundary condition 100
   initial condition 100
   flux limiters 123
   smart 123–124, 131–133
   van Leer 123–124
   geometric classification 100
   elliptic 106
   hyperbolic 100
   parabolic 102
   Heaviside function 101, 108, 110–111, 113
   hyperbolic PDE 100
   first order 100–101
   boundary condition (BC) 100
   initial condition (IC) 100
   main program 109
   ODE/MOL routine 109, 115
   second order 101
   boundary conditions 101
   expressed as first order 101–102
   initial conditions 101
   hyperbolic-parabolic PDE 107, 165
   Lagrangian coordinate 100–101, 113
   Laplace's equation 106
   Laplacian operator 104–105
   Cartesian coordinates 104
   cylindrical operator 105
   spherical coordinates 105
   LHS PDE time derivatives 68
   main program 77
   numerical graphical output 80–90
   mixed derivatives 99, 102
   Neumann boundary condition 101, 106, 153
   notation 99
   subscript 99, 101
   numerical integration
   parabolic PDE 102, 137–144
   ODE/MOL routine 144–149
   Peclet number 165–167, 172–173
   RHS PDE terms 90–94, 98
   main program 92–93

PDE *see also* one, two, three PDE (*cont.*)
    numerical graphical output 92–98
    Robin boundary condition 101–102, 106, 147–148, 153–155, 158
    second order hyperbolic 101
    boundary conditions 101
    expressed as first order 101–102
    initial conditions 101
    simultaneous PDEs *see also* two, three PDE models 166–167
    cocurrent 168
    countercurrent 168
    main program 168–174
    ODE/MOL routine 174–177
    numerical graphical output 177–180
    smart flux limiter 123–125, 131–136
    spherical coordinates 100
    step function 113
    traveling wave 100
    two point upwind 29–31, 46, 48, 77, 79, 108, 115–116, 121
    van Leer flux limiter 123–131, 155–163
    wave equation *see* second order hyperbolic
Peclet number 165–167, 172–173

R scientific computing system 2
    availabilty 2, 107
    [] matrix subscripting 6
    % number format 6
    d 6
    e 6
    f 6
    = replace 5, 7
    > greater than 11
    < less than 28
    && and operator 28
    # comment 5
    " " 6–7
    ; end of code 5, 7
    {} text brackets 6
    c numerical vector 11–12
    \n new line 6–7
    <<- return 11
    by 6
    cat 6
    deSolve 5, 7–8
    for 6
    function 5, 10–11
    if 10
    library 5, 7
    list 5, 7, 11–12
    lsodes 6, 8–9, 14, 22, 26, 40, 44, 109, 112
    ODE/MOL function 6–7, 10–11
    atol 6
    func 6
    maxord 6
    out 6
    rtol 6
    sparsetype 6
    y (IC) 6
    times 6
    matplot 7, 14–15
    matpoints 110
    matrix 6
    subscripting 6
    ncol 6, 14
    nrow 6, 14
    par 7
    persp 7, 16
    rep 6
    rm remove files 5, 7
    return 11–12
    seq 6
    from 6
    to 6
    by 6
    setwd 5, 7
    source 5, 7
    sprintf 6

Index 187

respiratory function 1, 58
    impaired respiratory function 1, 58
RHS PDE terms 90–94, 98
    main program 92–93
    numerical graphical output 92–98
Robin boundary condition 101–102, 106, 147–148, 153–155, 158

SARS-CoV-2 1
simultaneous PDEs *see also* two, three PDE models 166–167
    cocurrent 168
    countercurrent 168
    main program 168–174
    ODE/MOL routine 174–177
    numerical graphical output 177–180
source of neurological effects *see* neurolicical effects
smart flux limiter 123–124, 131–133
spatial derivative 1–3
spatial grid 8
spatiotemporal modeling 1–3
spherical coordinates 100
step function 113

three pde model
    blood to brain tissue transfer
    boundary conditions (BC) 44, 46–48, 55, 61–64
    cognitive impairment 67
    convection 40, 43, 46, 48
    dependent variables
    blood oxygen concentration 46
    brain tissue oxygen concentration 47
    neuron cell density 39, 47
    diffusivity 39, 42, 47
    discontinuity 41, 51
    finite difference (FD)
    three point centered 47, 49
    two point centered 47, 49
    two point upwind 46, 48
    graphical output 41, 45, 52–53
    initial conditions (IC) 39, 43
    main program 39
    mass transfer coefficient 40, 42
    neuron time derivative 39–40, 47, 49
    rate constant 39–40, 43, 47, 49, 53
    numerical output 40, 44–45, 50–51
    ODE/MOL routine 39–40, 44, 46–48
    spatial derivative 47
    spatial grid 40, 43
    time derivatives 39, 48–49
    time interval 40, 43
    time variation of the brain oxygen concentration 55
    velocity 40, 43, 46, 48
time derivative 1–3, 11
time interval 8
time variation of the blood oxygen concentration 55
    boundary condition 55, 57, 79
    constant 55, 57–59
    sine variation 55, 57–59
    concentration 55
    convection 55, 57
    diffusion 55
    diffusivity 55, 57, 79
    discontinuity 56, 80
    finite difference (FD)
    three point centered 77, 79
    two point centered 77, 79
    two point upwind 77, 79
    graphical output 56–58, 80–90
    lung recovery 58
    neuron rate constant 56–57, 78
    numerical output 60–61
    ODE/MOL routine 55, 59, 80
    mass transfer coefficient 55, 57, 79
    respiratory function recovery 58
    recovery 58

time variation of the blood oxygen concentration (*cont.*)
   supplemental oxygen 58
   time derivative vectors 78, 98
   velocity 55, 57, 79
traveling wave 100
two PDE model
   blood to brain tissue transfer 17–19, 29
   boundary conditions (BC) 17, 27–29, 31
   convection 17, 21, 24, 29
   dependent variables
   blood oxygen concentration 17
   brain tissue oxygen concentration 17–18
   diffusivity 17–19, 21, 24, 28
   discontinuity 23, 27
   finite difference (FD) 28–29
   order 30
   three point centered 30–31
   two point centered 30–31
   two point upwind 29–31
   graphical output 34–36
   initial conditions (IC) 17, 22, 25
   main program 21
   mass transfer coefficients 17, 21, 24
   numerical output 32–33
   ODE/MOL routine 22, 27
   spatial derivative 28
   spatial grid 22, 25
   time derivative 28, 32
   time interval 22, 25
   velocity 17, 21, 24, 29

upwind finite difference 29–31, 46, 48, 77, 79, 108, 115–116, 121

van Leer flux limiter 123–131, 155–163
velocity 1–3, 5, 7–8, 10, 14

wave equation *see* hyperbolic PDE, second order